Lecture Notes in Bioinformatics 8738

Subseries of Lecture Notes in Computer Science

François Fages   Carla Piazza (Eds.)

Formal Methods in Macro-Biology

First International Conference, FMMB 2014
Nouméa, New Caledonia, September 22-24, 2014
Proceedings

 Springer

Volume Editors

François Fages
Inria
Bâtiment 8, Domaine de Voluceau
78150 Rocquencourt, France
E-mail: francois.fages@inria.fr

Carla Piazza
University of Udine
Department of Mathematics and Computer Science
Via delle Scienze 206
33100 Udine, Italy
E-mail: carla.piazza@uniud.it

ISSN 0302-9743   e-ISSN 1611-3349
ISBN 978-3-319-10397-6   e-ISBN 978-3-319-10398-3
DOI 10.1007/978-3-319-10398-3
Springer Cham Heidelberg New York Dordrecht London

Library of Congress Control Number: 2014946072

LNCS Sublibrary: SL 8 – Bioinformatics

Typesetting: Camera-ready by author, data conversion by Scientific Publishing Services, Chennai, India

Printed on acid-free paper

Springer is part of Springer Science+Business Media (www.springer.com)

Preface

This volume contains the papers presented at FMMB 2014, the First International Conference on Formal Methods in Macro-Biology, held during September 21–23, 2014, in Noumea, New Caledonia.

Over the past decade, formal methods from computer science have been successfully applied in life sciences to decipher biological processes mostly at the molecular and cellular levels. Extending these methods to higher levels in systems biology, such as tissues, organs, but also populations and ecosystems is a challenging issue. In order to analyze such complex systems, temporal and spatial models able to represent a large number of components acting at different scales are required. Beyond ordinary and partial differential equation systems, powerful modeling languages, approximation techniques, and efficient algorithms need be designed, by involving experts from different disciplines, for tackling challenging macro-biology questions.

The aim of FMMB is to bring together researchers, developers, and students in theoretical computer science, applied mathematics, mathematical and computational biology, interested in studying the application of formal methods to the construction and analysis of models describing biological processes at both micro and macro levels. In order to achieve such a goal, we relied on 29 experts of disciplines related to macro-biology involved in the Program Committee.

In response to the call for papers, the Program Committee received 17 submissions and selected 10 of them to be included in the conference program. Each paper was revised on the average by three referees and the selection was based on originality, quality, and relevance for the conference. The scientific program consisted of papers on a wide variety of topics, including ecological systems, medical applications, logical frameworks, and –more in general– discrete continuous and hybrid models for the analysis of biological systems at macroscopic levels. The program also comprised seven invited talks describing challenging and successful experiences in such fields.

We would like to thank the people, institutions, and companies that contributed to the success of this first edition of FMMB. We gratefully acknowledge the financial support from the Government of New Caledonia, Inria, GdR BIM of CNRS, IRD, and the University of New Caledonia. We also wish to thank in advance Province Nord, Province Sud, Province des Îles, and Ministère des Outre-Mer, which, at the time of writing, were deciding whether to fund the conference.

We acknowledge Springer for publishing the proceedings of FMMB and the EasyChair management system for providing a useful platform for conference administration.

Finally, we would like to extend our special thanks to Teodor Knapik, as well as to the Organizing Committee, for his efforts, not only in making the local arrangements and organizing an attractive social program, but also for motivating and supporting us along all the decisional steps.

We hope this first edition of the conference will serve as a springboard for collaborations, projects, and for all the future editions of FMMB.

July 2014   François Fages
                                                                                                             Carla Piazza

Organization

Program Committee

Pierre Auger	Académie des Sciences and IRD, France
Pieter Collins	Maastricht University, The Netherlands
Thao Dang	Verimag, France
Finn Drablos	Norwegian University of Science and Technology, Norway
Saso Dzeroski	Jozef Stefan Institute, Slovenia
François Fages	Inria Paris-Rocquencourt, France (Co-chair)
Ofer Feinerman	Weizmann Institute, Israel
Ricard Gavaldà	Universitat Politècnica de Catalunya, Spain
Radu Grosu	Vienna Technical University, Austria
Katsumi Inoue	National Institute of Informatics, Japan
Steffen Klamt	Max Planck Institute Magdeburg, Germany
Nicolas Le Novère	Babraham Institute, UK
Pietro Liò	Cambridge University, UK
Pablo Marquet	Pontificia Universidad Católica de Chile
Annabelle McIver	Macquarie University, Australia
Istvan Miklos	Hungarian Academy of Sciences
Bud Mishra	New York University, USA
Satoru Miyano	University of Tokyo, Japan
Hélène Morlon	Ecole Normale Supérieure (Paris) and CNRS, France
Ion Petre	Åbo Akademi University, Finland
Carla Piazza	University of Udine, Italy (Co-chair)
Jean-Christophe Poggiale	Aix Marseille University, France
David Safranek	Masaryk University, Czech Republic
Daniel Stouffer	University of Canterbury, New Zealand
Gouhei Tanaka	University of Tokyo, Japan
Ps Thiagarajan	National University of Singapore
Denis Thieffry	ENS Paris, France
Jerzy Tiuryn	Warsaw University, Poland
Adelinde Uhrmacher	Universität Rostock, Germany

Organizing Committee

Didier Caucal	CNRS and Univeristy of Marne la Vallée, France
Teodor Knapik	University of New Caledonia (Chair)
Morgan Mangeas	IRD Montpellier, France

Additional Reviewers

Sepinoud Azimi   Gianvito Pio
Cristian Gratie   Sven Thiele
Md. Ariful Islam   Aneta Trajanov
Morgan Magnin

Sponsors

Government of New Caledonia
Inria - Institut National de Recherche en Informatique et en Automatique
GdR BIM of CNRS - GdR de Bioinformatique Moléculaire of CNRS
IRD - Institut de Recherche pour le Développement
UNC - Université de la Nouvelle-Calédonie

Invited Talks

Medical Cyber-Physical Systems:
The Heart Challenge

Radu Grosu[1,2]

[1] Vienna University of Technology
[2] Stony Brook University

Abstract. This talk discusses the opportunities and research challenges faced in the modeling, analysis and control of the human heart. Consisting of more than 4 billion communication nodes, interconnected through a very sophisticated communication structure, this ultimate cyber-physical system achieves with an astonishing reliability, the electric synchronization and the mechanical contraction of all of its nodes, in order to pump blood, during what is commonly known as a heart beat. However, even this cyber-physical system, engineered by billion years of evolution is fallible. Predicting and controlling its failure is a great challenge for our society.

Fast Enumeration of Smallest Metabolic Engineering Strategies in Genome-Scale Networks

Steffen Klamt

Max Planck Institute for Dynamics of Complex Technical Systems,
Magdeburg, Germany
klamt@mpi-magdeburg.mpg.de

One ultimate goal of metabolic network modeling is the rational modification of biochemical networks to optimize the bio-based production of certain compounds. Although several constraint-based optimization techniques have been proposed for this purpose, there is still a need for computational approaches allowing an effective systematic enumeration of efficient intervention strategies in large-scale metabolic networks.

Here we present the *MCSEnumerator* approach by which a large number of the smallest genetic intervention strategies (with fewest targets) can be readily computed in genome-scale metabolic models [1]. The algorithm builds upon an extended concept of Minimal Cut Sets (MCSs) which are minimal combinations of reaction (or gene) deletions leading to the fulfillment of a predefined intervention goal. It exploits the fact that smallest MCSs can be calculated as shortest elementary modes in a dual network and uses an improved procedure for shortest elementary-modes calculation. A broad spectrum of intervention problems can be formulated in a very convenient manner: one only needs to provide a description of the desired and undesired behaviors (flux distributions) by means of linear inequalities.

Realistic application examples demonstrate that our algorithm is able to list thousands of the most efficient intervention strategies for various intervention problems in genome-scale networks. We used *MCSEnumerator* to compute strain designs for growth-coupled synthesis of different products by heterotrophic as well as photoautotrophic organisms. We found numerous new engineering strategies partially requiring fewer interventions and guaranteeing higher product yields than reported previously.

In summary the presented approach can quickly calculate a large number of smallest metabolic engineering strategies with neither network size nor the number of required interventions posing major challenges. With its speed and the high flexibility in formulating intervention problems, we expect *MCSEnumerator* to become an important tool for Metabolic Engineering.

Moreover, the mathematical approach originally developed for finding interventions in biochemical reaction networks can be easily applied to any type of material flow networks, including those at macroscopic and ecosystems levels.

Reference

1. von Kamp, A., Klamt, S.: Enumeration of smallest intervention strategies in genome-scale metabolic networks. PLoS Computational Biology 10, e1003378 (2014)

Model-Checking in Systems Biology - From Micro to Macro

Pieter Collins

Department of Knowledge Engineering
Maastricht University

Abstract. In this paper, we will address some of the main challenges in the development of efficient, rigorous methods for model-checking the multiscale, highly uncertain dynamic systems arising in biological applications, and provide some theory and methods which address these. We will focus on three main aspects: 1) how to represent and compute with continuous data-types in a mathematically sound way, 2) how to handle modular and hierarchical systems, including abstraction of dynamics and system reduction, and 3) how to model uncertainties and handle the resulting models numerically. We will illustrate some of these issues and possible approaches for a model arising in cardiac electrophysiology.

Keywords: Hybrid automata, Numerical Tools, Analysis, Reachability, Verification, Electrophysiology.

Quantifying the Complexity
of "Complex" Ecological Networks

Daniel B. Stouffer

School of Biological Sciences
University of Canterbury
Christchurch, New Zealand
daniel.stouffer@canterbury.ac.nz

Abstract. Ecological networks are thought to be archetypal complex systems where the whole is far greater than the sum of its parts. Structurally, however, it remains unclear whether ecological networks are actually the product of unbridled complexity or if there are actually only a few ways to build an ecological network. In order to disentangle this problem, recent work has focused on developing new tools with which to quantify network complexity at a variety of scales—from the whole network down to individual species. Ultimately, such research will help us to better understand the way that species and species interactions are literally woven together in a larger network, and how that network acts in turn to place structural constraints on its constituent species.

Keywords: antagonistic networks, complex networks, ecological complexity, ecological networks, food webs, mutualistic networks, network motifs

Despite the fact that present-day ecosystems face threats, such as climate change or habitat loss, that permeate across entire communities [1], ecological understanding is built primarily from studies of one to a few species. Recognising this apparent disconnect, ecology has started to move toward a more holistic "network approach"—built upon robust analytical tools from applied mathematics, statistical physics, and the social sciences—that considers all species within an ecosystem and the collection of interactions between them such as who eats whom or which pollinators visit which flowers [1]. Overall, this network approach to ecology has strongly reinforced the idea that ecological networks are archetypal complex systems where the whole is far greater than the sum of its parts.

Despite the fact that empirical networks appear structurally complex and highly non-random, there are a number of statistical regularities that transcend the specifics of the different ecosystems [2, 6, 7, 10]. For example, food webs—the networks of predator-prey interactions in an ecosystem—show robust patterns on a global scale whether they come from environments such as deserts, lakes, streams, or rain forests [8]. One particularly intriguing result along these lines has demonstrated that certain network motifs [3, 10]—three species connected together by a specific set of interactions—appear more often in empirical food webs than would be expected at random and thus represent the simple "building blocks" of empirical food webs [10]. Furthermore, these motif patterns appear to be conserved over deep evolutionary time [9]. In contrast to the idea of ecological networks encapsulating unbridled complexity, results like these could well indicate that food webs are constrained by forces that allow species turnover—either via dispersal, speciation, or extinction—so long as it maintains a relatively constant topological structure. If this is true, there may actually be only a few viable ways to "build" an ecological network.

In this contribution, I will highlight novel methods to quantify the complexity of ecological networks using the concept of network motifs as a common thread. While doing so, I will pay particular attention to mechanisms that act as drivers of and limits to variation between species. Within a network context, such variation is of critical importance since a species' interactions have been directly linked to multiple emergent properties, such as that species' vulnerability to extinction [4]. In addition, I will discuss the ways in which a species-level perspective should help uncover the strategies that allow individual species to persist in the face of increasing threats to biodiversity, such as invasive species [5].

References

1. Bascompte, J.: Disentangling the web of life. Science 325(5939), 416–419 (2009)
2. Guimerà, R., Stouffer, D.B., Sales-Pardo, M., Leicht, E., Newman, M., Amaral, L.: Origin of compartmentalization in food webs. Ecology 91, 2941–2951 (2010)
3. Milo, R., Shen-Orr, S., Itzkovitz, S., Kashtan, N., Chklovskii, D., Alon, U.: Network motifs: Simple building blocks of complex networks. Science 298(5594), 824–827 (2002)
4. Saavedra, S., Stouffer, D.B., Uzzi, B., Bascompte, J.: Strong contributors to network persistence are the most vulnerable to extinction. Nature 478(7368), 233–235 (2011)
5. Sala, O.E., Chapin, F.S., Armesto, J.J., Berlow, E., Bloomfield, J., Dirzo, R., Huber-Sanwald, E., Huenneke, L.F., Jackson, R.B., Kinzig, A., Leemans, R., Lodge, D.M., Mooney, H.A., Oesterheld, M., Poff, N.L., Sykes, M.T., Walker, B.H., Walker, M., Wall, D.H.: Global biodiversity scenarios for the year 2100. Science 287(5459), 1770–1774 (2000)
6. Stouffer, D.B., Camacho, J., Amaral, L.A.N.: A robust measure of food web intervality. Proc. Natl. Acad. Sci. USA 103(50), 19015–19020 (2006)
7. Stouffer, D.B., Camacho, J., Guimerà, R., Ng, C.A., Amaral, L.A.N.: Quantitative patterns in the structure of model and empirical food webs. Ecology 86, 1301–1311 (2005)
8. Stouffer, D.B., Ng, C.A., Amaral, L.A.N.: Ecological engineering and sustainability: A new opportunity for chemical engineers. AIChE J. 54(12), 3040–3047 (2008)
9. Stouffer, D.B., Sales-Pardo, M., Sirer, M.I., Bascompte, J.: Evolutionary conservation of species' roles in food webs. Science 335(6075), 1489–1492 (2012)
10. Stouffer, D.B., Camacho, J., Jiang, W., Amaral, L.A.N.: Evidence for the existence of a robust pattern of prey selection in food webs. Proc. R. Soc. Lond. B 274(1621), 1931–1940 (2007)

Developing Quantitative Methods in Community Ecology: Predicting Species Abundances from Qualitative Web Interaction Data

Hugo Fort

Department of Physics, Faculty of Sciences, Universidad de la República,
Iguá 4225 Montevideo 11400
hugo@fisica.edu.uy

Abstract. Quantitative predictions of biodiversity of human-impacted ecological communities are crucial for their management. In the case of plant-pollinator mutualistic networks, despite the great progress in describing the interactions between plants and their pollinators, the capability of making quantitative predictions is still in its infancy. Furthermore, a general problem is the lack of measures or estimations of species abundances.

Here I propose a general method to estimate pollinator species abundances and their niche distribution from the available data, namely network interaction matrices. It works by transforming a plant-pollinator network into a competition model between pollinator species. Competition matrices were obtained from 'first principles' calculations, using qualitative interaction matrices compiled for a set including more than 40 plant-pollinator networks. This method is able to make accurate quantitative predictions for mutualistic networks spanning a broad geographic range. Specifically, the predicted biodiversity metrics for pollinators - species relative abundances, Shannon equitability and Gini-Simpson indices - agree quite well with those inferred from empirical counts of visits of pollinators to plants.

The importance of interspecific competition between pollinator species is a controversial and unresolved issue, considerable circumstantial evidence has accrued that competition between insects does occur, but a clear measure of its impact on their species abundances is still lacking. The present work contributed to fill this gap by quantifying the effect of competition between pollinators.

Particular applications could be to estimate the quantitative effects of removing a species from a community or to address the fate of populations of native organisms when foreign species are introduced to ecosystems far beyond their home range. This method also allows building a one-dimensional niche axis for pollinators in which clusters of generalists are separated by specialists thus rendering support to the theory of emergent neutrality.

Keywords: Quantitative ecology, species abundances, Lotka-Volterra competition.

Understanding How Biodiversity Is Distributed in Space and Time

Hélène Morlon

Institut de Biologie de l'Ecole Normale Supérieure, UMR 8197 CNRS, 46 rue d'Ulm,
75005 Paris, France
morlon@biologie.ens.fr
http://www.biologie.ens.fr/phyloeco/index.html

Abstract. Attention to biodiversity issues has been growing in the recent years. Despite the urgency of the problem, the development of a general theory of biodiversity is still underway. How can we develop such a theory? Two main approaches have dominated the field: the first approach has emphasized ecological controls of biodiversity, and has sought to explain static biodiversity patterns, often referred to as macroecological patterns; the second approach has emphasized historical controls of biodiversity, and has sought to explain more dynamic, evolutionary patterns of biodiversity, often referred to as macroevolutionary patterns. I will quickly summarize these approaches, focusing on the analytical and computational methods that have been used. Then, I will discuss how we can hope to integrate these two approaches to obtain a theory of biodiversity that better accounts for both historical and ecological factors.

Keywords: biodiversity theory, macroecology, macroevolution.

Inductive Process Modeling
for Learning the Dynamics of Biological Systems

Sašo Džeroski[1,2,3]

[1] Department of Knowledge Technologies, Jožef Stefan Institute
Jamova cesta 39, SI-1000 Ljubljana, Slovenia
[2] Jožef Stefan International Postgraduate School, Ljubljana, Slovenia
[3] Center of Excellence for Integrated Approaches to the Chemistry and Biology of
Proteins, Ljubljana, Slovenia
Saso.Dzeroski@ijs.si

Process-based models (PBMs) represent dynamic systems at two levels of abstraction. At the higher qualitative level, PBMs consist of entities and processes that describe the interactions between them. At the lower quantitative level, the models of individual processes provide details that allow the transformation of PBMs to systems of ODEs that can be used to precisely model the dynamics of the underlying system.

Inductive process modeling (IPM), a branch of computational scientific discovery, is concerned with the automated construction of PBMs from process-based modeling knowledge and measured data. The PB modeling knowledge describes classes of entities and processes in the modeling domain of discourse. This approach allows for constructing understandable PBMs, as well as the identification of both the structure and parameters of their underlying systems of ODEs.

IPM facilitates modular representation and reuse of domain knowledge. It has been successfully used to model biological systems at both the micro-level (i.e., to model cellular processes, such as endocytosis) and the macro-level (i.e., to model population dynamic processes in aquatic ecosystems, such as lakes and lagoons). We will introduce the task of inductive process modeling, describe some recent IPM approaches, and illustrate their use for modeling biological systems at both the micro- and macro- level.

Acknowledgments. S. Džeroski is supported by the European Commission through the projects MAESTRA - Learning from Massive, Incompletely annotated, and Structured Data (Grant no. 612944) and HBP - Human Brain Project (Grant no. 604102).

References

1. Atanasova, N., Todorovski, L., Dzeroski, S., Kompare, B.: Constructing a library of domain knowledge for automated modelling of aquatic ecosystems. Ecological Modelling 194, 14–36 (2006)

2. Atanasova, N., Dzeroski, S., Kompare, B., Todorovski, L., Gal, G.: Automated discovery of a model for dinoflagellate dynamics. Environmental Modelling and Software 26, 658–668 (2011)
3. Bridewell, W., Langley, P., Todorovski, L., Dzeroski, S.: Inductive process modeling. Machine Learning 71, 1–32 (2008)
4. Cerepnalkoski, D., Taskova, K., Todorovski, L., Atanasova, N., Dzeroski, S.: The influence of parameter fitting methods on model structure selection in automated modeling of aquatic ecosystems. Ecological Modelling 245, 136–165 (2012)
5. Cerepnalkoski, D.: Process-Based Models of Dynamical Systems: Representation and Induction. PhD thesis, Jozef Stefan International Postgraduate School, Ljubljana, Slovenia (2013)
6. Dzeroski, S., Todorovski, L.: Encoding and using domain knowledge on population dynamics for equation discovery. In: Magnani, L., Nersessian, N.J., Pizzi, C. (eds.) Logical and Computational Aspects of Model-Based Reasoning, pp. 227–247. Kluwer, Dordrecht (2002)
7. Dzeroski, S., Todorovski, L.: Equation discovery for systems biology: finding the structure and dynamics of biological networks from time course data. Current Opinion in Biotechnology 19, 360–368 (2008)
8. Langley, P., Sanchez, J., Todorovski, L., Dzeroski, S.: Inducing process models from continuous data. In: Proc. Nineteenth International Conference on Machine Learning, pp. 347–354. Morgan Kaufmann, San Francisco (2002)
9. Tanevski, J., Todorovski, L., Kalaidzidis, Y., Džeroski, S.: Inductive process modeling of rab5-rab7 conversion in endocytosis. In: Fürnkranz, J., Hüllermeier, E., Higuchi, T. (eds.) DS 2013. LNCS, vol. 8140, pp. 265–280. Springer, Heidelberg (2013)
10. Todorovski, L.: Using Domain Knowledge for Automated Modeling of Dynamic Systems with Equation Discovery. PhD Thesis, University of Ljubljana, Slovenia (2003)

Computing Longevity: Insights from Controls

Pietro Lió

Computer Laboratory, University of Cambridge, UK
pietro.lio@cl.cam.ac.uk

Abstract. There is a growing perception that medical treatment could be effective against aging although not as one intensive short time medication as we do with infections. It will require a precise, personalised knowledge of the genes and pathways that are perturbed during the progression to aging. Environmental factors, parental longevity and childhood are important predictors of exceptional longevity. Here we analyse molecular data (gene expression) from "healthy" controls of different age from several studies and we identify perturbations in key pathways affecting the susceptibility to several diseases. This work is exploratory and provide a useful test on existing data and methods for future studies.

Table of Contents

Model-Checking in Systems Biology - From Micro to Macro

Pieter Collins

Department of Knowledge Engineering
Maastricht University, The Netherlands

Abstract. In this paper, we will address some of the main challenges in the development of efficient, rigorous methods for model-checking the multiscale, highly uncertain dynamic systems arising in biological applications, and provide some theory and methods which address these. We will focus on three main aspects: 1) how to represent and compute with continuous data-types in a mathematically sound way, 2) how to handle modular and hierarchical systems, including abstraction of dynamics and system reduction, and 3) how to model uncertainties and handle the resulting models numerically. We will illustrate some of these issues and possible approaches for a model arising in cardiac electrophysiology.

Keywords: Hybrid automata, Numerical Tools, Analysis, Reachability, Verification, Electrophysiology.

1 Introduction

Formal methods have a long history in the study of computer systems. Although such systems may exhibit highly complicated interactions, they are discrete, so can be described and studied exactly, and mathematically rigorous results are readily obtained.

In systems biology, we are typically dealing with continuous state- and time-systems, which provide very different challenges. Aside from the need to handle continuous dynamics defined by differential equations, biological systems pose particular problems for a rigorous analysis. Such systems are typically highly complex; even a single cell contains hundreds of organelles and thousands of different chemical compounds and reaction pathways. Cells are not identical, but contain variations which means that even cells of the same type from the same organism behave differently. Further, many of of basic processes are not well understood; reaction pathways may be incomplete and reaction and diffusion rates are typically only known to coarse approximations.

Which formal methods may be used in systems biology to verify that a putative model of a dynamic system satisfies experimentally-observed properties, in this case, non-rigorous approaches usually suffice. However, formal methods may still be invaluable in highly complex multiscale systems, for which the accumulated effect of many heuristic approximations in the modelling and analysis process may significantly alter the behaviour of a non-rigorous simulation. In

F. Fages and C. Piazza (Eds.): FMMB 2014, LNBI 8738, pp. 1–22, 2014.

such cases, only by carefully controlling errors introduced in system reduction and abstraction, and in computation of the evolution, can reliability of the results be guaranteed.

Formal methods come into their own in cases where experimental validation of the prediction of a mathematical model is infeasible and correctness is paramount. In advanced medical devices, safety considerations may require a complete verification of the workings of the device, from control logic to mechanical properties, including integration with the patient's physiology. In personalised medicine, treatments regimes are designed based on knowledge of the patient's individual genotype and phenotype, and cannot be tested by large-scale clinical trials as in traditional therapies.

When developing formal methods for continuous systems, we need an appropriate theory of computation for continuous mathematics. In particular, since we are working with real numbers, which form an uncountable set, we cannot encode all numbers using a finite amount of data. There are a number of essentially equivalent approaches to computability theory for continuous mathematics [30], including Scott domain theory [26,31], type-two effectivity [59], recursive function theory [40] and realisability theory [8] and lambda calculus [57]. In this paper we will use type-two effectivity as a basis a computable type theory, which we believe is technically the simplest approach.

The main subject of this paper is the computability of many of the operations which will be needed to apply formal methods in macro-biology. We hope that work on computability and semantics will be of use in the development of practical tooling. We describe the basics of model-checking temporal logic formulae for continuous time systems, address abstraction and reduction issues relating to compositional systems, and provide a discussion of uncertainties in continuous systems. We also describe some of the challenges in their implementation, and mention some tools for rigorous analysis and model-checking continuous systems, but omit details of possible concrete numerical methods. We give a small case study on cardiac electrophysiology [58], but were unable to provide a validation of system properties using the methods implemented in our tool ARIADNE [6].

2   Computable Analysis

In order to use formal methods to give rigorous answers to questions about continuous systems, we need to first understand computation for continuous mathematics. Unlike discrete computation, in which mathematical objects can be described exactly, objects of continuous mathematics lie in uncountable sets, and so require an infinite amount of information to specify exactly. This leaves us with two obvious possibilities: we can insist on a finite theory of computation, but can then only work with countable subsets, or work with complete mathematical data types, but then need to consider computations on infinite data. It turns out we can develop a theory combining the strengths of both approaches: we formally work with infinite data streams, which greatly simplifies theoretical analysis, but ensure that useful information can be obtained from finite pieces of data, so the theory is practically relevant.

We begin by discussing computability theory for the real numbers in Section 2.1. This material is well known, see e.g. [12,47] We then extend the computability theory in Section 2.2 to general types in the formalism of [17], and provide fundamental logic and number types in Section 2.3 In Section 2.4 we describe types for sets, based on work of [57], and in Section 2.5 describe types of probability distributions [38] and random variables [56]. For a more detailed exposition of the material in this section, see [15].

2.1 Real Numbers

Since the real numbers are uncountable, any representation capable of describing *all* real numbers must necessarily use an infinite amount of data.

The decimal representation uses symbols -,.,0,1,...,9. Given a partial decimal expansion, e.g. $\pi = 3.14159\cdots$, we can deduce that $\pi \in [3.14159, 3.14160]$, or that $|\pi - 3.141595| \leq 5 \times 10^{-6}$. Hence useful information (in the form of interval bounds) can be extracted from the decimal expansion. However, it turns out that the decimal expansion is *not* appropriate as a data type of real numbers, since it is not always possible to compute the results of arithmetical operations. For if $x = 3 \times 0.33333\cdots$, we cannot determine whether $x < 1$ or $x > 1$, so cannot even output the first digit of the result!

For this reason, a better approach theoretically is to use a *signed-digit* representation, which includes symbols representing negative digits. The *binary* signed-digit representation uses a symbol $\bar{1}$ representing -1. For example, the number $3/4$ may be written in standard binary as $0.11000\cdots$ or $0.10111\cdots$, but in the signed-digit representation as $1.0\bar{1}000\cdots$ or $0.111\bar{1}\bar{1}\bar{1}\cdots$. It can be shown that the usual arithmetical operations $+, -, \times, \div$ are computable using the signed-digit representation.

The information provided by the signed-digit (or equivalent) representation is *not* sufficient to decide whether a number x is positive, since the sequence starting 0.000 may continue $0.00011\cdots$, in which case x is positive, $0.0000\bar{1}0\cdots$, in which case x is negative, or $0.000000\cdots$, in which case x is zero. Hence the tests $x > 0$ and $x \geq 0$ cannot be considered as functions $\mathbb{R} \to \mathbb{B}$, since the case $x = 0$ cannot be decided. However, if $x > 0$, then $x \geq 2^{-n}$ for some n, in which case we can prove $x > 0$ from $n + 1$ digits of the signed-digit representation. To avoid the inconvenience of partial functions, we use *three-valued* logic $\mathbb{T} = \{\mathsf{T}, \mathsf{F}, \uparrow\}$ where $\uparrow$ means "unknown" or "unknowable". Then we have a predicate $\mathbb{R} \to \mathbb{T}$ testing $x \gtrsim 0$ with value T if $x > 0$, F if $x < 0$, and $\uparrow$ if $x = 0$. Similarly, equality is undecidable, but inequality is verifiable, so we have a function $\neq \colon \mathbb{R} \times \mathbb{R} \to \{\mathsf{T}, \uparrow\}$

A crucial property of the signed-digit representation is that given a real number x, we can estimate x to any desired precision by a concrete rational number. In other words, given an accuracy $n \in \mathbb{N}$, we can find $q_n \in \mathbb{Q}$ such that $|x - q_n| \leq 2^{-n}$; this number can be chosen to have the form $q_n = m_n/2^n$ for some $m_n \in \mathbb{Z}$. The *Cauchy representation* of $\mathbb{R}$ is the representation by sequences of rationals q_n which are *strongly convergent* i.e. $|q_{n_1} - q_{n_2}| \leq 2^{-\min(n_1, n_2)}$. Given $|x - q_n| \leq 2^{-n}$, we clearly deduce that x lies in the interval $[q_n - 2^{-n}, q_n + 2^{-n}]$.

Conversely, given a stronly-convergent Cauchy sequence converging to x, we can compute a signed-digit expansion of x, so the Cauchy representation is equivalent to the signed-digit representation.

Use of Cauchy representation allows taking limits of strongly convergent sequences of real numbers. Suppose $\boldsymbol{x} \in \mathbb{R}^\omega$ is a sequence of real numbers such that $|x_m - x_n| \leq 2^{-m}$ whenever $m \leq n$. Take rationals $q_{n,k}$ such that $|q_{n,k} - x_n| \leq 2^{-k}$, and set $r_n = q_{n+1,n+2}$. Then if $m \leq n$, we have $|r_m - r_n| \leq |q_{m+1,m+2} - x_{m+1}| + |x_{m+1} - x_{n+1}| + |x_{n+1} - q_{n+1,n+2}| \leq 2^{-m-2} + 2^{-m-1} + 2^{-n-2} \leq 2^{-m}$. So $(r_n)_{n \in \mathbb{N}}$ is a strongly convergent Cauchy sequence, and clearly converges to $\lim_{n \to \infty} x_n$. The ability to take limits of strongly convergent Cauchy sequences of real numbers allows the definition of transcendental functions and the solution of differential and algebraic equations, amongst other operation.

As well as representations of the real numbers using infinite data streams, we can also define real numbers by symbolic formulae e.g. $\sin(2/3)$. The most general finitely-representable class of real numbers are the *computable reals*, which are described by computer programs outputting the symbols of the signed digit representation.

Note that we *do not* use (double-precision) floating-point numbers as a real number type! These numbers are insufficient even to represent decimals exactly e.g. 3.1 is rounded to $\frac{31}{10} + \frac{1}{5 \cdot 2^{51}}$. Floating-point numbers are an efficient data type for calculations, but should not be directly available to users as a specification type.

2.2 Computable Types

We have seen that real number representation and computation requires the use of infinite data *streams* Σ^ω over some finite alphabet Σ. Data streams are mapped to mathematical objects by *representations*, which are partial surjective functions $\delta : \Sigma^\omega \twoheadrightarrow X$. A δ-*name* of $x \in X$ is a sequence $\boldsymbol{w}$ such that $\delta(\boldsymbol{w}) = x$. From the theory of type-two effectivity [59], computations are performed by non-halting Turing machines, and define partial functions $\eta : \Sigma^\omega \twoheadrightarrow \Sigma^\omega$. Given representations $\delta_X : \Sigma^\omega \twoheadrightarrow X$ and $\delta_Y : \Sigma^\omega \twoheadrightarrow Y$, a function $f : X \to Y$ is *computable* if there is a machine-computable function $\eta : \Sigma^\omega \twoheadrightarrow \Sigma^\omega$ such that for all $\boldsymbol{w} \in \mathrm{dom}(\delta_X)$, $\delta_Y(\eta(\boldsymbol{w})) = f(\delta_X(\boldsymbol{w}))$.

Representations δ_1, δ_2 of X are *equivalent* if there are machine-computable functions γ_{12} and γ_{21} such that $\delta_2 \circ \gamma_{21} = \delta_1$ and $\delta_1 \circ \gamma_{12} = \delta$. A *computable type* is a pair $(X, [\delta])$, where δ is an equivalence-class of representations.

In order to ensure that useful information can be extracted from a finite piece of a name $\boldsymbol{w}$ of x, representations should be *admissible* with respect to some topology on X. It can be shown that any computable function is (sequentially) continuous. This means that the category of computable types with computable functions as morphisms is a full subcategory of the category of computable types with continuous functions.

Of critical importance is that computable types form a *Cartesian closed category*, which means they are a model of intuitionistic/constructive type theory. In particular, there is a terminal type $\mathbb{I}$, canonical product types $\mathbb{X}_1 \times \mathbb{X}_2$ and exponential types $\mathbb{Y}^{\mathbb{X}}$. Elements of $\mathbb{X}$ are naturally associated with functions $\mathbb{I} \to \mathbb{X}$, so a computable element is one with a machine-computable name, and form a countable subset of all elements. The product type $\mathbb{X}_1 \times \mathbb{X}_2$ naturally describes pairs (x_1, x_2) with $x_1 \in \mathbb{X}_1$ and $x_2 \in \mathbb{X}_2$. The type $\mathbb{Y}^{\mathbb{X}}$ consists of the (sequentially) continuous functions from $\mathbb{X}$ to $\mathbb{Y}$, via a computable *evaluation* operator $\varepsilon : \mathbb{Y}^{\mathbb{X}} \times \mathbb{X} \to \mathbb{Y}$, so is denoted by $\mathbb{X} \to \mathbb{Y}$ or $\mathcal{C}(\mathbb{X}; \mathbb{Y})$.

There is a natural computable equivalence between $\mathbb{Z}^{(\mathbb{X} \times \mathbb{Y})}$ and $(\mathbb{Z}^{\mathbb{Y}})^{\mathbb{X}}$ i.e. between $\mathbb{X} \times \mathbb{Y} \to \mathbb{Z}$ and $\mathbb{X} \to (\mathbb{Y} \to \mathbb{Z})$. This equivalence can be used to prove computability of objects defined by function types, since a function $f : \mathbb{X} \to (\mathbb{Z}^{\mathbb{Y}})$ is computable if the function $\tilde{f} : \mathbb{X} \times \mathbb{Y} \to \mathbb{Z}$ defined by $\tilde{f}(x, y) = \varepsilon(f(x), y)$ is computable.

If objects y of type $\mathbb{Y}$ are defined in terms of objects of type $\mathbb{X}$ as $y = f(x)$, we shall sometimes say that y is computable (from x) to mean that f is a computable function. Given a type $\mathbb{X} = (X, [\delta])$, we shall henceforth write $x \in \mathbb{X}$ to mean that x is an element of the set X described by a δ-name.

2.3   Fundamental Types

We can now place our real numbers within the framework of computable type theory. The real numbers are associated with a canonical type $\mathbb{R}$ such that arithmetic operations are computable, the sign of a nonzero number can be extracted, and the limits of strongly convergent Cauchy sequences can be computed; see [8]. The *signed-digit* representation is one of the representations for this type.

Since comparison of real numbers is undecidable, we need Kleene's three-valued logic type $\mathbb{T}$ with values $\{\mathsf{T}, \mathsf{F}, \uparrow\}$. A concrete representation of this type is $\delta(0^*11 \cdots) = \mathsf{T}$, $\delta(0^*10 \cdots) = \mathsf{F}$ and $\delta(0^\omega) = \uparrow$; note that T, F can be seen as the result of a finite computation, and $\uparrow$ as a non-terminating computation. The positivity test $x \gtrsim 0$ is a computable function $\mathbb{R} \to \mathbb{T}$. The usual booleans $\mathbb{B} = \{\mathsf{T}, \mathsf{F}\}$ are a subtype of $\mathbb{T}$ describing the results of decidable predicates. The *Sierpinski* type is the two element subtype $\{\mathsf{T}, \uparrow\}$ describing the results of verifiable predicates. Finite disjunction $\wedge : \mathbb{S} \times \mathbb{S} \to \mathbb{S}$ and countable conjunction $\bigvee : \mathbb{S}^\omega \to \mathbb{S}$ are both computable.

2.4   Logic and Sets

Since the power set of $\mathbb{R}$ has higher than continuum cardinality, we cannot handle arbitrary sets. However, the usual classes of open, closed and compact sets can be handled, and we also have a type of *overt* sets. These types are related to logic via the Sierpinski type.

A proposition p on a type $\mathbb{X}$ is verifiable if, and only if, it is a computable function $\mathbb{X} \to \mathbb{S}$ where $\mathbb{S}$ is the Sierpinski type. A set U is open if its characteristic function χ_U is a continuous function X to the Sierpinski space. Hence we identify the type $\mathcal{O}(\mathbb{X})$ of open subsets of $\mathbb{X}$ with $\mathbb{S}^{\mathbb{X}}$, so verifiable predicates are true on computable open sets. Similarly, closed sets $\mathcal{A}(\mathbb{X})$ are sets for which membership is falsifiable.

It is straightforward to show that the standard operations on sets are computable, including finite intersection of $\mathcal{O}(\mathbb{X})^* \to \mathcal{O}(\mathbb{X})$, countable union $\mathcal{O}(\mathbb{X})^\omega \to \mathcal{O}(\mathbb{X})$, complement $\mathcal{O}(\mathbb{X}) \leftrightarrow \mathcal{A}(\mathbb{X})$, and preimage $\mathcal{C}(\mathbb{X}, \mathbb{Y}) \times \mathcal{O}(\mathbb{Y}) \to \mathcal{O}(\mathbb{X})$: $(f, V) \mapsto f^{-1}(V)$. A proof that the preimage is computable is simple: for $f \in \mathcal{C}(\mathbb{X}, \mathbb{Y})$ and $V \in \mathcal{O}(\mathbb{Y})$, the operator $(f, V) \mapsto f^{-1}(V)$ can be computed since $x \in f^{-1}(V) \iff f(x) \in V \iff \chi_V(f(x))$.

For a set S, verification of the statement $\forall x \in S$, $p(x)$ is equivalent to testing $S \subset p^{-1}(\mathsf{T})$. Suppose S is a set such that any condition $S \subset U$ for open U is verifiable, so a computable function $\mathcal{O}(\mathbb{X}) \to \mathbb{S}$. Then whenever $(U_n)_{n \in \mathbb{N}}$ are open sets such that $\bigcup_{n=0}^{\infty} U_n = U$, continuity implies $S \subset \bigcup_{n=0}^{N-1} U_n$ for some N. The open cover $\{U_n\}_{n \in \mathbb{N}}$ of S thus has a finite subcover, so S is (countably) compact. We therefore identify the type of compact subsets $\mathcal{K}(\mathbb{X})$ of $\mathbb{X}$ with the subtype of $\mathbb{S}^{\mathcal{O}(\mathbb{X})}$ via the subset relation $S_\subset : \mathcal{O}(\mathbb{X}) \to \mathbb{S}$, and note that $S \subset (U_1 \cap U_2) \iff (S \subset U_1) \wedge (S \subset U_2)$. Similarly, we can consider when we can verify statements of the form $\exists x \in S$, $p(x)$, corresponding to predicates of the form $S \bowtie U$ for open U, where $\bowtie$ denotes 'intersects'. This gives rise to a type known as the type of *overt* sets, $\mathcal{V}(\mathbb{X})$ as the subtype of $\mathbb{S}^{\mathcal{O}(\mathbb{X})}$ satisfying $S \bowtie (U_1 \cup U_2) \iff (S \bowtie U_1) \vee (S \bowtie U_2)$. Note that both $\mathcal{K}(\mathbb{X})$ and $\mathcal{V}(\mathbb{X})$ are subtypes of $(\mathbb{X} \to \mathbb{S}) \to \mathbb{S}$.

Computable operations on compact and overt sets include countable union $\mathcal{V}^\omega(\mathbb{X}) \to \mathcal{V}(\mathbb{X})$ and finite union $\mathcal{K}^*(\mathbb{X}) \to \mathcal{K}(\mathbb{X})$, intersection $\mathcal{V}(\mathbb{X}) \times \mathcal{O}(\mathbb{X}) \mapsto \mathcal{V}(\mathbb{X})$ and $\mathcal{K}(\mathbb{X}) \times \mathcal{A}(\mathbb{X}) \mapsto \mathcal{K}(\mathbb{X})$, singleton $x \mapsto \{x\}$ for $\mathbb{X} \to \mathcal{V}(\mathbb{X})$ and $\mathbb{X} \to \mathcal{K}(\mathbb{X})$, and image $(f, S) \mapsto f(S)$ for $\mathcal{C}(\mathbb{X}, \mathbb{Y}) \times \mathcal{V}(\mathbb{X}) \to \mathcal{V}(\mathbb{Y})$ and $\mathcal{C}(\mathbb{X}, \mathbb{Y}) \times \mathcal{K}(\mathbb{X}) \to \mathcal{K}(\mathbb{Y})$.

A set R is *regular* if $\partial R^\circ = \emptyset$; such a set is naturally defined by taking $\chi_R : \mathbb{X} \to \mathbb{T}$, yielding a type $\mathcal{R}(\mathbb{X}) \equiv \mathbb{T}^{\mathbb{X}}$. A set S is *located* if both $S \subset U$ and $S \bowtie U$ are verifiable for any open U. The predicate $S \subset R$ can be decided whenever $S \subset R^\circ$, or if $S \not\subset \overline{R}$, the latter being equivalent to $S \bowtie (\mathbb{X} \setminus R)^\circ$. We therefore associate the type of located sets $\mathcal{L}$ as a subtype of $(\mathbb{X} \to \mathbb{T}) \to \mathbb{T}$.

A type $\mathbb{X}$ is *effectively Hausdorff* if inequality is verifiable, and this implies that any compact set is effectively closed; i.e. that there is a computable function $\mathcal{K}(\mathbb{X}) \to \mathcal{A}(\mathbb{X})$, given by $x \notin C \iff C \subset \{y \mid y \neq x\}$.

2.5   Probability

Traditional probability theory is based on σ-algebras and measurable functions rather than topology and continuity. This approach is not amenable to a computable theory, since general Borel subsets cannot be effectively approximated either from below or above. However, open closed sets can be approximated from below, and it is therefore reasonable to also be able to approximate the measure of such sets from below.

Just as the Sierpinski type forms the basis of topology, so the *lower halfspace* $\mathbb{H}$ forms the basis of measure theory and probability. The type $\mathbb{H}$ is the type of positive real numbers (and infinity), represented by increasing sequences of rational numbers.

A *valuation* [38] on $\mathbb{X}$ is a continuous function $\nu : \mathcal{O}(\mathbb{X}) \to \mathbb{H}$ satisfying $\nu(U_1 \cup U_2) + \nu(U_1 \cap U_2) = \nu(U_1) + \nu(U_2)$, In other words, valuations only give lower-measures, and only of open sets. This is natural since open sets are those for which we can approximate from below. A probability measure on $\mathbb{X}$ is a valuation P such that $P(\mathbb{X}) = 1$.

Valuations are naturally equivalent to integrals of lower-semicontinuous functions $\psi : \mathbb{X} \to \mathbb{H}$ by $\int_{\mathbb{X}} \psi \, d\nu = \sup\{\sum_{m=1}^{n} (p_m - p_{m-1}) \nu(\psi^{-1}(p_m, \infty]) \mid (p_0, \ldots, p_n) \in \mathbb{Q}^*, \, 0 = p_0 < p_1 < \cdots < p_n\}$. The integral of a valuation $\nu : (\mathbb{X} \to \mathbb{S}) \to \mathbb{H}$ is an additive functional $(\mathbb{X} \to \mathbb{H}) \to \mathbb{H}$. Conversely, from such a functional, we can define a valuation by $\nu(U) = \int \chi_U$. If $\psi : \mathbb{X} \to \mathbb{R}$ is continuous and bounded, then we can define $\int_{\mathbb{X}} \psi \, d\nu$ in $\mathbb{R}$.

Random variables can also be defined following [56,22]. We take underlying probability space $\Omega = \{0, 1\}^\omega$ with $P(\{w \in \Omega \mid w_i = c_i, \, i = 0, \ldots, k-1\}) = 2^{-k}$ for any word c. A continuous random variable taking values in a space $\mathbb{E}$ is a continuous function $X : \Omega \to \mathbb{E}$, with distribution $\Pr(X \in U) = P(X^{-1}(U))$. The space of measurable random variables is the completion of $\mathcal{C}(\Omega, \mathbb{E})$ under the *Fan metric* $d(X, Y) = \inf\{\epsilon > 0 \mid \Pr(d(X, Y) > \epsilon) < \epsilon\}$. The expectation of a continuous real-valued random variable is $E(X) = \int_\Omega X(\omega) \, dP(\omega)$, and expectations of measurable random variables are defined by taking limits since $E(\omega \mapsto d(X(\omega), Y(\omega))) \leq 2d(X, Y)$.

3   Model-Checking Continuous Systems

In this section we consider consider deterministic continuous-time systems. Such systems are typically described by ordinary differential equations, with a bounded set of possible initial conditions. We are interested in questions of whether a system satisfies certain properties, such as every trajectory starting in a set W of initial states remains in a set S of safe states.

We will show that it is possible to rigorously solve model-checking problems, but only in the case that the property under consideration is *robustly* true or false. A basic approach is to discretise time by solving the differential equation over a finite time interval, and then to perform model-checking on the resulting continuous-space system.

In Section 3.1, we state computability of the solution of differential equations; this result is standard, and was given in [55]. Verifiable semantics of linear temporal logic formulae are described in Section 3.2 based on work in [16,21].

3.1   Differential Equations

The simplest class of continuous systems are those described by ordinary differential equations $\dot{x} = f(x)$ for $x \in \mathbb{R}^n$. Under mild assumptions on f, the solution $\xi(t)$ starting at point x_0 exists and is unique for all $t \in \mathbb{R}^+ := [0, \infty)$. The following theorem is from [55].

Theorem 1 (Ruohonen). *Suppose the solution of the differential equation $\dot{x} = f(x)$ starting at x_0 is unique. Then the solution ξ is computable as a function $\xi : \mathbb{R}^+ \to \mathbb{R}^n$ given names of $f : \mathbb{R}^n \to \mathbb{R}^n$ and $x_0 : \mathbb{R}^n$.*

In particular, if f is a (locally) Lipschitz function, then all solutions are computable.

The *flow* of the differential equation is the function $\phi : \mathbb{R}^n \times \mathbb{R}^+ \to \mathbb{R}^n$ satisfying $\dot{\phi}(x,t) = f(\phi(x,t))$ and $\phi(x,0) = x$. By Ruohonen's theorem, ϕ is computable from f if all solutions exists and are unique. It is easy to show that ϕ satisfies the flow-condition $\phi(x,t_1 + t_2) = \phi(\phi(x,t_1),t_2)$. Continuous systems on more general spaces can be defined directly by a function $\phi : \mathbb{X} \times \mathbb{R}^+ \to \mathbb{X}$ satisfying $\phi(x,t_1 + t_2) = \phi(\phi(x,t_1),t_2)$ and $\phi(x,0) = x$.

3.2   Model-Checking Linear Temporal Logic

In formal methods, we often wish to validate system properties expressed in terms of temporal logic. In *linear temporal logic* [4], we make statements about *all* trajectories of the system defined in terms of temporal operators, including $\bigcirc$ (next), $\Diamond$ (eventually) and $\square$ (always). For simplicity, we will restrict to properties which can be expressed using the modalities *always* $\square$ and *eventually* $\Diamond$, notably

- The *target* property $\forall\Diamond\Psi$.
- The *safety* property $\forall\square\Psi$.
- The *recurrence* property $\forall\square\Diamond\Psi$.
- The *stabilisation* property $\forall\Diamond\square\Psi$.

We consider a continuous-time system S defined by a semiflow $\phi : \mathbb{X} \to (\mathbb{R}^+ \to \mathbb{X})$ with initial set $X_0 : \mathcal{K}(\mathbb{X})$, and temporal logic formulae based on an *atomic proposition* $\Psi : \mathbb{X} \to \mathbb{S}$.

The *target property* $S \models \forall\Diamond\Psi$ translates to

$$\forall x_0 \in X_0, \ \forall \xi \in \phi(x_0), \ \exists t \in \mathbb{R}^+, \ \Psi(\xi(t)).$$

This is verifiable since the set of all trajectories $\phi(X_0) = \{\phi(x_0) \mid x_0 \in X_0\}$ is a compact subset of $\mathbb{X}^{[0,\infty)}$, and *path formula* $\Diamond\Psi$ defines a set of valid trajectories $\{\eta \in \mathbb{X}^{[0,\infty)} \mid \exists t \in \mathbb{R}^+, \ \Psi(\eta(t))\}$ which is open in $\mathbb{X}^{[0,\infty)}$. since $S(X_0) \in \mathcal{K}(\mathcal{C}(\mathbb{R}^+,\mathbb{X}))$, and since $\{\eta \in \mathcal{C}(\mathbb{R}^+,\mathbb{X}) \mid \Psi(\eta(t))\} \in \mathcal{O}(\mathcal{C}(\mathbb{R}^+,\mathbb{X}))$.

If we now consider the safety property $S \models \forall\square\Psi$ we have a problem, since the path formula defines a set of paths

$$\square\Psi = \{\eta : \mathbb{R}^+ \to \mathbb{X} \mid \forall t \in [0,\infty), \Psi(\eta(t))\}$$

which is not open due to the universal quantification over a non-compact set. To prove safety, we need to use the semiflow property $\phi(x,t_1 + t_2) = \phi(\phi(x,t_1),t_2)$. Fix $T > 0$ (any T will do). Suppose we can choose open U and compact K with $U \subset K$ and verify

$$\phi(T,K) \subset U, \ \phi([0,T],K) \subset \Psi^{-1}(\mathsf{T}), \ X_0 \subset U.$$

Then for any $x_0 \in X_0$ and $t \in \mathbb{R}^+$, we can write $t = nT + r$ and show $\phi(nT + r, x) \in \phi(nT+r, K) \subset \phi_r(\phi_T^n(K)) \subset \phi_r(K) \subset U$. Note that the condition $U \subset K$

cannot be verified, but must be valid by construction of U and K; this is possible as long as the type $\mathbb{X}$ is *effectively locally compact*. Conversely, we showed in [16] that if safety is *robust*, then it may be verified as above. However, there are non-robust cases for which the system is safe, but cannot be proved to be so.

For the recurrence property, we have $\forall\square\lozenge\Phi \equiv \forall\square\forall\lozenge\Psi$, so recurrence is verifiable if we can verify safety and target properties.

For the stabilisation property $\forall\lozenge\square\Psi$, we showed in [21] that $\forall\lozenge\square\Psi \equiv \forall\lozenge\forall\square(\Psi\vee \forall\lozenge\square\Psi)$. Hence the set of initial points robustly satisfying the stabilisation property can be computed as

$$\bigvee_{n=0}^{\infty}\Theta_n \quad \text{where} \quad \Theta_0 = \forall\lozenge\forall\square\Psi; \quad \Theta_{n+1} = \forall\lozenge\forall\square(\Psi\vee\Theta_n).$$

Here, the operator $\forall\square(\Psi\vee\Theta_n)$ should be interpreted as the *robust* safety operator as above.

4   Compositional Systems

A macroscopic system is composed of many such subprocesses, each with its own internal state x_i, which interact via relationships $u_i = g_i(y_1, \ldots, y_m)$ between the external inputs u_i and outputs y_j. We are ultimately interested in checking some temporal logic property about the behaviour of global quantities derived from the outputs of the individual components. In particular, the internal dynamics is unobservable and the details are not of experimental interest. For simplicity, we only consider the case of two components here.

In Section 4.1 we give standard definitions of systems theory, which can be found in [51]. In Section 4.2 we describe the *regular expression representation* of an input-output system; for more details see [19]. In Section 4.3, we outline a general approach to reducing the complexity of a system while retaining guarantees on the error, motivated by work on the rigorous reduction of infinite-dimensional systems by Galerkin projection [61,23].

4.1   External Behaviour

We consider a standard state-space model S of an input-output system

$$\dot{x} = f(x,u), \quad y = h(x), \quad x(0) \in X_0 \tag{1}$$

where $u \in \mathbb{U}$, $x \in \mathbb{X}$ and $y \in \mathbb{Y}$ are the *input*, *state* and *output*, respectively. In principle, we are only interested in the *external* behaviour, which is given by an *input-output* mapping

$$\mathcal{M}[S] : \mathbb{U}^{[0,\infty)} \rightrightarrows \mathbb{Y}^{[0,\infty)} : \mu \mapsto \{\eta \mid \exists \xi \in \mathbb{X}^{[0,\infty)}, \ \dot{\xi} = f(\xi,\mu) \wedge \eta = h(\xi) \wedge \xi(0) \in W\}.$$

The input-output map $M = \mathcal{M}[S]$ is *causal*, which means

$$\mu_1|_{[0,T]} = \mu_2|_{[0,T]} \implies M(\mu_1)|_{[0,T]} = M(\mu_2)|_{[0,T]}.$$

Consider input spaces $\mathbb{U}_{1,2}$ and output spaces $\mathbb{Y}_{1,2}$ such that $\mathbb{U}_1 = \mathbb{U}_0 \times \mathbb{Y}_2$ and $\mathbb{U}_2 = \mathbb{U}_0 \times \mathbb{Y}_1$. Given systems $S_{1,2}$, define the *parallel composition* $S_1 \| S_2$ to be the system with input space $\mathbb{U}_0$, state space $\mathbb{X}_1 \times \mathbb{X}_2$ and output space $\mathbb{Y}_1 \times \mathbb{Y}_2$ such that

$$(\dot{x}_1, \dot{x}_2) = f_{1,2}((x_1, x_2), u_0) = (f_1(x_1, (u_0, h(x_2))), f_2(x_2, (u_0, h(x_1)))); \quad (y_1, y_2) = (h_1(x_1), h_2(x_2)).$$

Given input-output maps M_1, M_2, we define the parallel composition as

$$M_1 \| M_2 : \mu_0 \mapsto \{(\eta_1, \eta_2) \mid \eta_1 \in M_1(\mu_0, \eta_2) \wedge \eta_2 \in M_2(\mu_0, \eta_1)\}.$$

It is straightforward to show that

$$\mathcal{M}[S_1 \| S_2] = \mathcal{M}[S_1] \| \mathcal{M}[S_2].$$

We now consider two problems in the study of compositional systems. The first is of *abstraction*, or the representation of the input-output behaviour itself. It can be shown that the data provided by the input-output map as an element of the type

$$(\mathbb{R}^+ \to \mathbb{U}) \to \mathcal{K}(\mathbb{R}^+ \to \mathbb{Y})$$

is *insufficient* to prove safety and stabilisation properties, and hence a different representation is needed. The second problem is that of *reduction*, or simplification of the state-space representation.

4.2   Regular Expression Representation

We now consider how to represent the external behaviour of a continuous-time state-space model in such a way that robust temporal logic properties of the original can still be proved. There are many motivations for providing a state-space independent descriptions, notably to allow different concrete state-space models to be compared and substituted for each other.

We start be considering the situation for discrete input-output automata, which are systems with finite input, state, and output spaces described by

$$x_{n+1} = f(x_n, u_n), \quad y_n = h(x_n), \quad x_0 \in X_0.$$

Given an input sequence $\boldsymbol{u}$ and initial state, there is a unique output sequence $\boldsymbol{y}$. The external behaviour can be expressed by the set of infinite sequences of inputs and generated outputs $(y_0, u_0, y_1, u_1, y_2, \ldots)$.

It can be shown that the set of possible input/output sequences is an ω-*regular language* which can be described by a *regular expression* [50]. Further, an equivalent finite-state automaton can be constructed from a regular expression describing the external behaviour, so regular expressions provide a canonical state-space independent way of describing the external behaviour.

Finite regular expressions define finite words, and include all literals, and are closed under *catenation* $R_1 R_2$, *alternation* $R_1 | R_2$ and *Kleene star* R^* are finite regular expression. Regular ω-expressions define infinite sequences and are defined recursively: If R is a regular expression and S, S_1, S_2 are regular ω-expressions, then the *catenation* RS, the *alternation* $S_1|S_2$ and the *sequentialisation* R^ω are regular ω-expressions. We will also require that our ω-expressions are closed sets; the simplest way of doing this is to require that the expression R^*S also includes R^ω.

For continuous-time systems, we can obtain a finite-state automaton description by discretising the state space, and spaces of input and output *functions*. For simplicity, we assume that inputs, states and outputs are uniformly bounded, so that $\mathbb{U}$, $\mathbb{X}$ and $\mathbb{Y}$ can be effectively taken to be compact. We also assume that inputs and outputs are Lipschitz with constant L; this is used to obtain compactness of closed balls in the supremum norm on continuous functions. This is no restriction, since if inputs and states remain bounded, we can bound both $y = h(x)$ and $\dot{y} = h'(x)\dot{x} = h'(x)f(x, u)$.

The state space $\mathbb{X}$ is discretised into finitely many (possibly overlapping) closed compact sets $\widehat{X}$. In order to discretise the inputs and outputs, we fix a time $T > 0$ and discretise the space of bounded Lipschitz functions $\mathbb{U}^{[0,T]}$ and $\mathbb{Y}^{[0,T]}$ into closed compact sets $\widehat{\mathcal{U}}$ and $\widehat{\mathcal{Y}}$. Given $x(0) \in \hat{x}$ and $\mu|_{[0,T]} \in \hat{\mu}$, we can compute $\{\xi : [0, T] \to \mathbb{X} \mid \xi(0) \in \hat{x} \wedge (\exists \mu \in \hat{\mu}, \ \dot{\xi} = f(\xi, \mu))\}$ and hence the output signals $\eta : [0, T] \to \mathbb{Y}$. We then discretise the set of all $\xi(T) \in \mathbb{X}$ and $\eta \in \mathbb{Y}^{[0,T]}$. The resulting system is a discrete input-output automaton with states $\widehat{X}$, input space $\widehat{\mathcal{U}}$ and output space $\widehat{\mathcal{Y}}$, so can be described by a regular expression over $\widehat{\mathcal{U}}, \widehat{\mathcal{Y}}$. For any sequence in $\widehat{\mathcal{U}}^\omega$ containing some input $\mu \in \mathbb{U}^{[0,\infty)}$, the resulting output sequence in $\widehat{\mathcal{U}}^\omega$ contains all possible system outputs $\eta \in \mathbb{Y}^{[0,\infty)}$. By taking arbitrarily fine discretisations of $\mathbb{U}^{[0,T]}$, $\mathbb{X}$ and $\mathbb{Y}^{[0,T]}$, we can approximate the behaviour to any desired accuracy.

There are many concrete ways of discretising the space of inputs and/or outputs; direct possibilities include approximation by step-functions or piecewise-linear functions, though these will typically have nontrivial overlaps. Other possibilities are to discretise the space $[0, T] \to \mathbb{R}$ first transforming to Fourier space or expressing functions in a wavelet basis; note that the Fourier series of L-Lipschitz functions satisfies $|a_n|, |b_n| \sim 1/n$.

We can now define the *regular expression representation* of an input-output map.

Definition 1. *The* regular expression representation *of an input-output map consists of a sequence of tuples $(\widehat{\mathcal{U}}_i, \widehat{\mathcal{Y}}_i, R_i)$ such that $\widehat{\mathcal{U}}_i, \widehat{\mathcal{Y}}_i$ are discretisations of Lipschitz functions in $\mathbb{U}^{[0,T]}, \mathbb{Y}^{[0,T]}$ by closed, compact sets, and R_i is a regular expression over $\widehat{\mathcal{U}}_i \times \widehat{\mathcal{Y}}_i$ such that any sequence admitted by R_{i+1} is a refinement of a sequence admitted by R_i, and that $\bigcup (\bigcap_{n=0}^{\infty} L(R_n)) = \text{graph}(M)$.*

It can be shown that *any* property which can be verified from a state-space representation of a system S can be verified using the regular-expression representation of the input-output map.

Theorem 2. *Let $S = (F, H, W)$ be a non-blocking input-output system such that $F : \mathbb{X} \times \mathbb{U} \to \mathcal{K}(\mathbb{X})$, $H : \mathbb{X} \to \mathcal{K}(\mathbb{Y})$ and $W \in \mathcal{K}(\mathbb{X})$. Then any property of S*

verifiable from (F, H, W) is also verifiable from a regular expression representation of the input-output map of S.

We note that this theorem only covers properties which can be verified using the description of F, H as continuous functions, some properties such as existence of fixed-points require effective differentiability.

We immediately obtain verifiability of properties of closed-loop systems

Corollary 1. *The target safety, recurrence and stabilisation properties for the parallel composition of S_1 with another system S_2 are verifiable from the regular expression representations of S_1 and S_2.*

In practise, we may not actually wish to use the regular expression representation to express the behaviour of a subsystem, but instead use a description which is closer to the original. Possibilities include representations as piecewise-constant derivative systems or timed automata, which have simple dynamics, to use a low dimensional state-space as described in the next section, or simply to discretise the state space. However, as long as the information provided by the regular expression representation can be extracted from a class of abstractions, the class is sufficient to prove any topological property.

4.3   System Reduction

A detailed approach to subsystem modelling often yields a model containing effects which have little impact on the external behaviour. The goal of system reduction is to create a simplified model with approximately the same input-output behaviour.

There are many ways of constructing *reduced order* component models, which usually take into account special structures of the system, such as linearity or time-scale separation. Most approaches approximate the differential input-output system (1) by another system

$$\dot{\tilde{x}}(t) = \tilde{f}(\tilde{x}(t), u(t)); \quad y(t) = \tilde{h}(\tilde{x}(t))$$

with the same input and output variables u, y, but state variables $\tilde{x} \in \widetilde{\mathbb{X}}$. Typically, x and $\tilde{x}$ are related by a projection $\tilde{x} = \pi(x)$ for $\pi : \mathbb{X} \to \widetilde{\mathbb{X}}$.

The reduced component model only provides an *approximation* of the external behaviour of the original model. In order to perform a rigorous analysis of the system dynamics, we need to control the errors introduced by the reduction. We first need to find a compact subset A_X of X which we can guarantee contains the true state x. We then show that the derivative of the projection $\pi(x)$ satisfies

$$\frac{d}{dt}(\pi \circ x) = \pi'(x)\dot{x} = \pi'(x)f(x, u).$$

and hence taking $\tilde{x} = \pi(x)$ we have

$$\dot{\tilde{x}} \in \widetilde{F}(\tilde{x}) := \{\pi'(x)f(x, u) \mid x \in \pi^{-1}(\tilde{x}) \cap A_X\}.$$

The resulting system is a *differential inclusion* for the projection $\tilde{x}$.

Given an approximate reduction $\dot{\tilde{x}} = \tilde{f}(\tilde{x}, u)$, we can alternatively write

$$\dot{\tilde{x}}(t) = \tilde{f}(\tilde{x}(t), u(t)) + \tilde{e}(t); \quad \tilde{e}(t) \in \widetilde{E}(\tilde{x}(t), u(t))$$

where the error set $\widetilde{E}$ is given by

$$\widetilde{E}(\tilde{x}, u) = \{\pi'(x)f(x, u) - \tilde{f}(\pi(x), u) \mid x \in \pi^{-1}(\tilde{x}) \cap A_X\}.$$

Finally, the output is given by

$$y \in \widetilde{H}(\tilde{x}) := \{h(x) \mid x \in \pi^{-1}(\tilde{x}) \cap A_X\}.$$

In many cases, we will have $h(x_1) = h(x_2)$ whenever $\pi(x_1) = \pi(x_2)$, in which case $\widetilde{H}$ reduces to a single-valued function $\tilde{h}$.

As long as $\pi : \mathbb{X} \to \widetilde{\mathbb{X}}$ is computable with computable derivatives, and $A_X : \mathcal{K}(\mathbb{X})$ is computable, then the set-valued function $\widetilde{F} : \widetilde{\mathbb{X}} \times \mathbb{U} \to \mathcal{K}(\widetilde{\mathbb{X}})$ is computable; note $\pi^{-1}(\tilde{x})$ is computable as a closed subset of $\mathbb{X}$.

The approach outlined above is only a skeleton of a method to perform model-order reduction. Firstly, an appropriate reduction $\pi : \mathbb{X} \to \widetilde{\mathbb{X}}$ and reduced dynamics $\tilde{x} \approx \tilde{f}(\tilde{x}, u)$ need to be found; this is highly dependent on the system under consideration, but can be done using existing approximative methods. One typical method is to find an (almost) invariant submanifold $\widetilde{X} = M \subset \mathbb{X}$. We note that for reduction methods which based on invariant manifolds, it suffices to compute an *approximate* functional parametrisation. It is more convenient to use a parametrisation of the manifold rather than a grid-based representation as in [33].

Secondly, an attractor A_X for the unreduced state x needs to be found. Whether a set A_X is attracting also depends on the input, so we typically need to assume that u lies in some compact set B_U. If A_X is defined by $a(x) \leq 0$, then A_X is invariant if $\nabla a(x) \cdot f(x, u) \leq 0$ whenever $a(x) = 0$ and $u \in B_U$. Since the error set $\widetilde{E}$ depends on A_X, the tighter A_X, the better the approximation of the resulting system. Hence it may be necessary to use an *assume-guarantee* approach, where successively narrowing B_U allows for smaller errors in the reduced system. It may also be necessary to use different reductions for different parts of the state space X. The resulting system is *hybrid* with switching occurring if the state estimate $\tilde{x}$ leaves the domain of validity of a given reduction, and possibly requiring re-inflation of the reduced state to full dimension. Switching from a reduced mode is also necessary if the input leaves its pre-specified bounds.

Thirdly, the reduced system is no longer a differential equation, but a nondeterministic *differential inclusion* $\dot{\tilde{x}} \in \widetilde{F}(\tilde{x}, u)$. The practical solution of differential inclusions is very difficult, and will be discussed in Section 5.

5   Uncertain Systems

Uncertain systems in continuous time can be written in the form

$$\dot{x}(t) = f(x(t), v(t)) \tag{2}$$

where $v(\cdot) \in \mathcal{V} \subset \mathbb{V}^{[0,\infty)}$ is a space of *noise* signals. Unlike the case of inputs $u(t)$,the noise is *not* under user control, and only partial information abouts its values is know.

We consider two classes of uncertain systems, namely *nondeterministic* systems for which the noise $v(t)$ is constrained to lie in some set V, and *stochastic* systems, for which the noise is a random variable. In particular we will consider *differential inclusions*

$$\dot{x} \in F(x) = \overline{\mathrm{conv}}\{f(x,v) \mid v \in V\}$$

and *stochastic differential equations*

$$dx(t) = f(x(t))dt + G(x(t))\, dW(t),$$

for which $v(t) = dW(t)/dt$ is a (formal) derivative of Brownian motion.

Although stochastic differential equations are currently more commonly used as models of uncertain systems, they in principle are based on an *exact* knowledge of the noise distribution which is unavailable in practise. Further, as we have seen in Section 4.3, differential inclusions arise naturally in model-order reduction, since the uncertainties are not independent but can be bounded. Other models of nondeterministic systems are possible, such as *imprecise* stochastic models, where the noise distribution is not known exactly, and combinations of nondeterministic noise $v(t) \in V$ and stochastic noise $v(t) = dW(t)/dt$.

The material on differential inclusions is contained in Section 5.1. An algorithm for computing arbitrarily accurate approximations to reachable sets was given in [52], and suffices to show computability; a direct proof is given in [20]. Computability of stochastic differential equations follows from a careful effectivisation of standard existsence and uniqueness results [29].

5.1   Differential Inclusions

A differential inclusion is a continuous time system of the form

$$\dot{x} \in F(x); \qquad F : \mathbb{R}^n \rightrightarrows \mathbb{R}^n. \tag{3}$$

A function $\xi : \mathbb{R}^+ \to \mathbb{R}^n$ is a solution if ξ for almost every $t \in \mathbb{R}^+$, ξ is differentiable at t and $\dot{\xi}(t) \in F(\xi(t))$.

The relationship between (3) and (2) for $v \in V$ is not entirely straightforward. It can be shown that if $f(x,V)$ is convex for all x, then for any solution $x(\cdot)$ of $\dot{x} \in f(x,V)$, there exists a continuous function $v : \mathbb{R}^+ \to V$ such that $\dot{x}(t) = f(x(t), v(t))$ for almost every v. Further, if $v_n(\cdot)$ is a sequence of continuous functions such that the sequence of solutions $x_n(t) = f(x_n(t), v_n(t))$ converges uniformly, then the limit x_∞ solves $\dot{x}_\infty \in \overline{\mathrm{conv}}(f(x_\infty, V))$. This suggests that the noisy differential equation $\dot{x} = f(x,v)$ should be modelled by the differential inclusion $\dot{x} \in \overline{\mathrm{conv}}(f(x,V))$. We note that the mapping $X \mapsto \overline{\mathrm{conv}}(X)$ is computable $\mathcal{L}(\mathbb{X}) \to \mathcal{L}(\mathbb{X})$. We henceforth assume $F : \mathbb{R}^n \rightrightarrows \mathbb{R}^n$ has closed, convex values.

Ideally, we aim to compute the mapping

$$\Phi_F : \mathbb{R}^n \rightrightarrows (\mathbb{R}^+ \to \mathbb{R}^n) : \Phi_F(x_0) = \{\xi : \mathbb{R}^+ \to \mathbb{R}^n \mid \xi(0) = x_0 \wedge \dot{\xi}(t) \in F(\xi(t)) \text{ a.e.}\}$$

which takes x_0 to the set of trajectories starting at x_0. However, for many applications it often suffices to compute the map

$$\mathbb{R}^n \times \mathbb{R}^+ \rightrightarrows \mathbb{R}^n : (x_0, t) \mapsto \{\xi(t) : \mathbb{R}^+ \to \mathbb{R}^n \mid \xi(0) = x_0 \wedge \dot{\xi}(\tau) \in F(\xi(\tau)) \text{ a.e.}\}$$

which gives the set of reachable points from a given starting point x_0 at a given time t without describing the intermediate trajectories.

In order to prove properties of *all* trajectories, we will need to consider differential inclusions $F : \mathbb{R}^n \to \mathcal{K}(\mathbb{R}^n)$. To ensure trajectories do not escape to infinity, we require the *bounded growth* condition that for some constant K, $|y| \leq K(1 + |x|)$ whenever $F(x) \ni y$. In order to prove existential properties, we need to consider differential inclusions where $F : \mathbb{R}^n \to \mathcal{V}(\mathbb{R}^n)$, and need a local *one-sided Lipschitz* condition, that for every $x_1, x_2 \in \mathbb{R}^n$ with $|x_1|, |x_2| \leq N$, and $y_1 \in F(x_1)$, there exists $y_2 \in F(x_1)$ such that $(x_1 - x_2) \cdot (y_1 - y_2) \leq L|x_1 - x_2|^2$.

The following theorems are in [20]; classical existence and uniqueness results are given in [3,24], explicit computations in [52] and semicontinuity results in [32].

Theorem 3. *Let F be a locally one-sided-Lipschitz lower-semicontinuous multivalued function with closed convex values. Then the solution operator $F \mapsto \Phi_F$ of the initial value problem $\dot{x} \in F(x); \ x(0) = x_0$ is computable as a function*

$$\mathcal{C}(\mathbb{R}^n, \mathcal{V}(\mathbb{R}^n)) \to \mathbb{R}^n \to \mathcal{V}(\mathcal{C}(\mathbb{R}^+, \mathbb{R}^n)).$$

Theorem 4. *Let F be an upper-semicontinuous multivalued function with compact convex values and bounded growth. Then the solution operator $F \mapsto \Phi_F$ of the initial value problem $\dot{x} \in F(x); \ x(0) = x_0$ is computable as a function*

$$\mathcal{C}(\mathbb{R}^n, \mathcal{K}(\mathbb{R}^n)) \to \mathbb{R}^n \to \mathcal{K}(\mathcal{C}(\mathbb{R}^+, \mathbb{R}^n)).$$

Practical methods have been implemented in a number of tools. For large V, the solution set grows rapidly, and coarse low-order approximations are most appropriate [11,5]. For small perturbations higher-order methods of [60,62] can provide highly-accurate solutions. The ellipsoidal methods of [41] combine low-order set representations with high-order temporal approximations.

5.2   Stochastic differential Equations

Stochastic processes of the form

$$\dot{X}(t) = f(X(t), V(t))$$

where V is a computable random variable with values in $\mathcal{C}(\mathbb{R}^+, \mathbb{V})$ can be solved directly pathwise,

$$\dot{X}(\omega, t) = f(X(\omega, t), V(\omega, t)).$$

The result is a random variable taking values in $\mathcal{C}(\mathbb{R}^+, \mathbb{X})$.

However, the most interesting case is a *stochastic differential equation* defined in terms of the Wiener process $W(\cdot)$ by

$$dx(t) = f(x(t))dt + G(x(t))dW(t).$$

Stochastic differential equations occur when the noise is uncorrelated with scale-free increments; in one dimension this means $W(t_1) - W(t_0) \sim N(0, t_1 - t_0)$ and is independent of W_0 for $t_1 > t_0$.

By effectivising the usual existence and uniqueness results for stochastic differential equations [18] based on the stochastic integral

$$\int_0^T X(t)\, dW(t),$$

we can prove the following:

Theorem 5. *The solution of the stochastic differential equation*

$$dx(t) = f(x(t))dt + G(x(t))dW(t)$$

where $f : \mathbb{R}^n \to \mathbb{R}^n$ and $G : \mathbb{R}^n \to \mathbb{R}^{n \times m}$ is computable as a random variable X taking values in $\mathcal{C}(\mathbb{R}^+, \mathbb{X})$. The equation may be interpreted using either the Itō or Stratonovich stochastic calculus.

We note that if V_n is a sequence of smooth processes such that $V_n \to W$ almost everywhere, in the limit, we obtain the *Stratonovich* form of the stochastic differential equation $dx(t) = f(x(t))dt + G(x(t)) \circ dW(t)$. There is some evidence to suggest that the Stratonovich interpretation is more suitable for physical systems [46].

Properties of stochastic processes can be expressed using probabilistic temporal logic. Finite-time properties are readily computable, but safety properties typically have probability 0.

The computability results show that it is possible to compute the solution of a stochastic differential equation as a random variable over sample paths, and effectively give rigorous lower bounds for the measure of any open set of paths. However, the computational difficulty of performing such calculations is high, and to the best of our knowledge such methods have not yet been implemented. The usual way of numerically solving stochastic differential equations is by performing multiple simulations in an attempt to estimate the probabilities. A detailed, but still non-rigorous, method for estimating probability densities of sample paths and reachable sets was given in [42]. An alternative approach for diffusion processes which have a continuous density function is to consider the Fokker-Planck partial differential equation e.g. [9].

6   Cardiac Myocyte Models

The electrophysiology of cardiac myocytes provides an interesting testing ground for the development of numerical implementations of the operations described in this paper. The key quantity is the *membrane potential* of a cell, which is the electric potential across the cell wall, with a physiological range of roughly $-80mV$ to $+40mv$. The potential is determined by the motion of charged sodium Na^+, potassium K^+, calcium Ca^{2+} and (to a lesser extent) chloride Cl^- ions into and out of the cell. The transport of ions across the cell membrane is through various *channels*, each of which gives rise to a current I. Transport may be passive or active, in the latter case allowing ions to move against a concentration gradient or electrostatic potential, but requiring input of energy. The opening and closing of channels is governed by a number of factors, notably the membrane potential itself, and by signalling molecules.

The basic equation for the current through a channel of type i is

$$I_i = g_i(V - E_i)$$

where g_i is a time-dependent *gating* variable and E_i is the reverse potential of the channel. A simple gating model is the *voltage-gating* model of Hodgekin and Huxley [37]. A similar model is the *slow-inward sodium current* used in a simple model of Tran et al. [58], and is given by

$$I(t) = \bar{g}d(t)f(t)(V(t) - E) \quad .$$

where the gating variables satisfy

$$\dot{d} = \alpha_d(V)(1 - d) - \beta_d(V)d; \quad \dot{f} = \alpha_f(1 - f) - \beta_f f$$

and the functions α_d, β_d, α_f and β_f are sigmoidal functions depending only on V. More complicated models consider the mechanism of channel opening and closing, described by a Markov model with multiple state variables. A detailed electrophysiological model is given in [36].

It is possible to isolate single cells in the laboratory and study their behaviour. In typical experiments, cells are subjected to an externally periodically-varying voltage $V_{stim}(t)$, typically in the form of a short pulse. Since only the membrane potential is of interest for the electrophysiological properties, we have a single-input, single-output system. The membrane potential for the model of [58] is shown in Figure 6. In vivo, cells are coupled, and stimuli travel across the surface of the heart or through specialised Purkinje fibres.

Ideally, we would like to prove that the output of a myocyte model does not exhibit any arrhythmias, such as *early afterdepolarisations*. A first challenge in formal methods for cardiac electrophysiology is to provide characterisations of normal behaviour and of arrhythmias in terms of temporal logic.

Even though the single-cell model of [58] only has four state variable, analysis by rigorous methods is extremely challenging. The system exhibits multiple time-scales, with rapid depolarisation due to the action potential followed by a plateau, and later depolarisation. Further, the form of the gating functions $\alpha_{d,f}, \beta_{d,f}$ in the model are complicated and make extensive use of exponential functions. This means that rigorous methods for the solution of differential equations may need to take small steps, and we were unable to integrate the

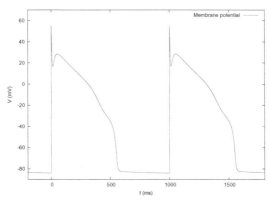

Fig. 1. Membrane potential for the Tran [2009] myocyte model

system using Taylor-series based integration methods in our tool ARIADNE [1]. We have considered simplifying the sigmoidal functions used for the gating by using splines [14]. Dealing with the multiple time scales may be possible by using straightforward adaptive methods, but may require specialised methods for stiff systems, which are currently not available for rigorous integrators.

Other challenges include the analysis of more complicated Markov gating models. In particular, we expect the behaviour of these models to be very well-described by a model with many fewer variables, making this a good candidate for the study of model-order reduction methods. Since many different models exist in the literature for the same ion channel, we would like to compare the input-output behaviour, ideally as described by discretisation and regular expressions. Further, models parameters are hard to determine experimentally, and many models are over-determined, and different cells have different characteristics, meaning that a careful analysis should take into account uncertainties in the model. Finally, moving from single-cell analysis to tissue and whole-heart analysis will require a combination of reduction and abstraction techniques.

7   Existing Tooling for Rigorous Numerics and Model-Checking

While basic methods for the analysis of well-conditioned deterministic systems exist, tooling for many of the more sophisticated algorithms still needs to be developed. A good base for this work are existing packages for rigorous numerical methods and model checking.

Multiple-precision rounded arithmetic is implemented by the MPFR library [27], and can be used as a base for interval arithmetic and exact real arithmetic, such as the iRRAM library [48]. Differential equations can be rigorously solved using AWA [43], VNODE [49], COSY Infinity [45], and the CAPD Library [44]. Methods for global analysis of dynamical systems, similar to model-checking technology is implemented in GAIO [25]

There are a large number of tools available for model-checking hybrid systems, including D/DT [2], Checkmate —citecheckmate2003, HSolver [53], Phaver [28], and Flow* [13]. The tool ARIADNE [6] aims at providing a complete functionality for model-checking nonlinear continuous-time and hybrid systems, and includes sub-modules for rigorous interval arithmetic, linear algebra, solution of algebraic and differential equations, and feasibility and optimisation problems. Work is in progress in implementing methods for differential inclusions, and long-term goals include support for ill-conditioned systems, model-order reduction, and stochastic systems.

Techniques based on formal methods from computer science have also been used for the non-rigorous analysis of biological systems. Typically, the continuous dynamics is approximated, either by non-rigorous numerical integration, simplification to a class of function more amenable to analysis, such as piecewise-affine systems, or both. While such approaches can provide useful scientific results, especially if the approximations made are accurate, the lack of rigour is theoretically unsatisfying and may cause problems in practise: if a tool suggests that a model does not satisfy experimentally-observed properties, is this a failure of the model or the tool? Work combining discrete model-checking with approximate analysis of the continuous dynamics includes work on cardiac myocytes [35,34] and yeast cell cycle models using linear temporal logic [54]. A rigorous approach based on rectangular partitions [10] is implemented in the Genetic Network Analyser [39] and BioDiVine [7], but these methods are incomplete in the sense that not all robust properties are provable. An interesting direction for future work would be to couple these tools with rigorous tools for reduction and analysis of continuous dynamics.

8   Concluding Remarks

In this paper we have described a framework for model-checking the complex, multi-component, highly uncertain systems arising in systems biology. We have focused on computability theory, which allows us to determine what is *possible* to achieve by rigorous methods, and also to study appropriate data representations and semantics of computable operations. The development of practical general, rigorous, efficient and usable implementations of most of the methods described here is a much greater challenge, but one which we believe it is important for the formal methods community to address. The work in this paper shows that such methods are theoretically possible, even for the complex, multiscale, and highly uncertain system models arising in biological applications. Existing tooling for rigorous numerical computation can provide a foundation for more advanced functionality for system analysis. We hope that in the future, tools based on rigorous numerical methods will gradually phase out tools using heuristic model reduction and approximative numerics.

References

1. Ariadne: An open tool for hybrid systems analysis, http://fsv.dimi.uniud.it/ariadne/

2. Asarin, E., Dang, T., Maler, O.: The **d/dt** tool for verification of hybrid systems. In: Brinksma, E., Larsen, K.G. (eds.) CAV 2002. LNCS, vol. 2404, pp. 365–370. Springer, Heidelberg (2002)

3. Aubin, J.P., Cellina, A.: Differential inclusions. Grundlehren der Mathematischen Wissenschaften, vol. 264. Springer, Berlin (1984)

4. Baier, C., Katoen, J.P.: Principles of model checking. MIT Press, Cambridge (2008), with a foreword by Kim Guldstrand Larsen

5. Baier, R., Gerdts, M.: A computational method for non-convex reachable sets using optimal control. In: Proceedings of the European Control Conference, pp. 97–102 (2009)

6. Balluchi, A., Casagrande, A., Collins, P., Ferrari, A., Villa, T., Sangiovanni-Vincentelli, A.L.: Ariadne: a framework for reachability analysis of hybrid automata. In: Proceedings of the 17th International Symposium on the Mathematical Theory of Networks and Systems, Kyoto, Japan, July 24-28, pp. 1269–1267 (2006)

7. Barnat, J., Brim, L., Cerna, I., Drazan, S., Fabrikova, J., Lanik, J., Safranek, D., Ma, H.: BioDiVinE: A framework for parallel analysis of biological models. In: Computational Models for Cell Processes, pp. 31–45 (2009)

8. Bauer, A.: The Realizability Approach to Computable Analysis and Topology. Ph.D. thesis, Carnegie Mellon University (2000)

9. Bect, J., Baili, H., Fleury, G.: Generalized Fokker-Planck equation for piecewise-diffusion processes with boundary hitting resets. In: Proceedings of the 17th International Symposium on the Mathematical Theory of Networks and Systems, Kyoto, Japan, July 24-28 (2006)

10. Belta, C., Habets, L.: Controlling a class of nonlinear systems on rectangles. IEEE Trans. Autom. Control 51(11), 1749–1759 (2006)

11. Beyn, W.J., Rieger, J.: Numerical fixed grid methods for differential inclusions. Computing 81(1), 91–106 (2007)

12. Boehm, H.J., Cartwright, R., Riggle, M., O'Donnell, M.: Exact real arithmetic: a case study in higher order programming. In: Proceedings of the ACM Symposium on Lisp and Functional Programming, pp. 162–173 (1986)

13. Chen, X., Ábrahám, E., Sankaranarayanan, S.: Flow*: An analyzer for non-linear hybrid systems. In: Sharygina, N., Veith, H. (eds.) CAV 2013. LNCS, vol. 8044, pp. 258–263. Springer, Heidelberg (2013)

14. Clerx, M., Collins, P.: Reducing run-times of excitable cell models by replacing computationally expensive functions with splines. In: Proceedings of the International Symposium on the Mathematical Theory of Networks and Systems (2014)

15. Collins, P.: Computable analysis with applications to dynamic systems. Tech. rep., Centrum Wiskunde en Informatica, Amsterdam, cWI Report MAC-1002 (2010)

16. Collins, P.: Optimal semicomputable approximations to reachable and invariant sets. Theory Comput. Syst. 41(1), 33–48 (2007)

17. Collins, P.: Computable types for dynamic systems. In: Proceedings of the Fifth Conference on Computability in Europe (2009)

18. Collins, P.: Computable stochastic processes (in preparation, 2014)

19. Collins, P.: Input-output representations for verifying safety properties. In: Proceedings of the International Symposium on the Mathematical Theory of Networks and Systems (2014)

20. Collins, P., Graça, D.: Effective computability of solutions of differential inclusions — the ten thousand monkeys approach. J. Universal Comput. Sci. 15(6), 1162–1185 (2009)

21. Collins, P., Zapreev, I.S.: Computable semantics for CTL* on discrete-time and continuous-space dynamic systems. Int. J. Found. Comput. Sci. 22(4), 801–821 (2011)
22. Coquand, T., Spitters, B.: Integrals and valuations. J. Logic Analysis 1(3), 1–22 (2009)
23. Day, S., Junge, O., Mischaikow, K.: A rigorous numerical method for the global analysis of infinite-dimensional discrete dynamical systems. SIAM J. Appl. Dyn. Syst. 3(2), 117–160 (2004)
24. Deimling, K.: Multivalued differential equations. de Gruyter Series in Nonlinear Analysis and Applications, vol. 1. Walter de Gruyter & Co., Berlin (1992)
25. Dellnitz, M., Froyland, G., Junge, O.: The algorithms behind GAIO-set oriented numerical methods for dynamical systems. In: Ergodic Theory, Analysis, and Efficient Simulation of Dynamical Systems, pp. 145–174, 805–807. Springer (2001)
26. Edalat, A.: Domains for computation in mathematics, physics and exact real arithmetic. Bull. Symbolic Logic 151(4), 401–452 (1997)
27. Fousse, L., Hanrot, G., Lefèvre, V., Pélissier, P., Zimmermann, P.: MPFR: A multiple-precision binary floating-point library with correct rounding. ACM Trans. Math. Softw. 33(2) (2007), http://www.mpfr.org/
28. Frehse, G., et al.: SpaceEx: Scalable verification of hybrid systems. In: Gopalakrishnan, G., Qadeer, S. (eds.) CAV 2011. LNCS, vol. 6806, pp. 379–395. Springer, Heidelberg (2011)
29. Friedman, A.: Stochastic Differential Equations. Dover (1975)
30. Geuvers, H., Niqui, M., Spitters, B., Wiedijk, F.: Constructive analysis, types and exact real numbers. Math. Structures Comp. Sci. 17(1), 3–36 (2007)
31. Gierz, G., Hofmann, K., Keimel, K., Lawson, J., Mislove, M., Scott, D.: Continuous lattices and domains. Cambridge University Press (2003)
32. Goebel, R., Teel, A.R.: Solutions to hybrid inclusions via set and graphical convergence with stability theory applications. Automatica J. IFAC 42(4), 573–587 (2006)
33. Gorban, A., Karlin, I., Zinovyev, A.: Invariant grids: Method of complexity reduction in reaction networks. Complexus 2004–2005 (2), 110–127 (2005)
34. Grosu, R., Batt, G., Fenton, F.H., Glimm, J., Le Guernic, C., Smolka, S.A., Bartocci, E.: From cardiac cells to genetic regulatory networks. In: Gopalakrishnan, G., Qadeer, S. (eds.) CAV 2011. LNCS, vol. 6806, pp. 396–411. Springer, Heidelberg (2011)
35. Grosu, R., Smolka, S.A., Corradini, F., Wasilewska, A., Entcheva, E., Bartocci, E.: Learning and detecting emergent behavior in networks of cardiac myocytes. Commun. ACM 52(3), 97–105 (2009)
36. Heijman, J., Volders, P.G., Westra, R.L., Rudy, Y.: Local control of beta-adrenergic stimulation: Effects on ventricular myocyte electrophysiology and ca(2+)-transient. Journal of Molecular and Cellular Cardiology (2011)
37. Hodgkin, A.L., Huxley, A.F.: A quantitative description of membrane current and its application to conduction and excitation in nerve. The Journal of Physiology 117(4), 500–544 (1952)
38. Jones, C., Plotkin, G.: A probabilistic powerdomain of evaluations. In: Proceedings of the Fourth Annual Symposium on Logic in Computer Science, pp. 186–195 (1989), http://dl.acm.org/citation.cfm?id=77350.77370
39. de Jong, H., Geiselmann, J., Hernandez, C., Page, M.: Genetic Network Analyzer: Qualitative simulation of genetic regulatory networks. Mathematical Biology 66(2), 261–300 (2004)

40. Kleene, S.C.: Mathematical Logic (1967)
41. Kurzhanski, A., Valyi, I.: Ellipsoidal calculus for estimation and control. In: Systems and Control: Foundations and Applications, Birkhäuser (1997)
42. Kushner, H., Yin, G.G.: Stochastic Approximation and Recursive Algorithms and Applications. Stochastic Modelling and Applied Probability, vol. 35. Springer (2003)
43. Lohner, R.J.: AWA–software for the computation of guaranteed bounds for solutions of ordinary initial value problems. Tech. rep., Institut fur Angewandte Mathematik (2006)
44. Mrozek, M., et al.: CAPD Library (2007), http://capd.ii.uj.edu.pl/
45. Makino, K., Berz, M.: COSY INFINITY Version 9. Nuclear Instruments and Methods A558, 346–350 (2006)
46. Mannella, R., McClintock, P.V.E.: Itô versus Stratonovic: 30 years later. Fluctuation Noise Lett. 11(1) (2012)
47. Menissier-Morain, V.: Arbitrary precision real arithmetic: design and algorithms. J. Logic Algebraic Programming 64, 13–39 (2005)
48. Müller, N.T.: The iRRAM: Exact arithmetic in C++. In: Blank, J., Brattka, V., Hertling, P. (eds.) CCA 2000. LNCS, vol. 2064, pp. 222–252. Springer, Heidelberg (2001), http://irram.uni-trier.de/
49. Nedialkov, N.: VNODE-LP. Tech. rep., cAS-06-06-NN (2006)
50. Perrin, D., Pin, J.E.: Infinite Words: Automata, Semigroups, Logic and Games. Elsevier (2004)
51. Polderman, J.W., Willems, J.C.: ntroduction to mathematical systems theory: A behavioral approach. Texts in Applied Mathematics, vol. 26. Springer, New York (1998)
52. Puri, A., Varaiya, P., Borkar, V.: Epsilon-approximation of differential inclusions. In: Alur, R., Sontag, E.D., Henzinger, T.A. (eds.) HS 1995. LNCS, vol. 1066, pp. 362–376. Springer, Heidelberg (1996)
53. Ratschan, S., She, Z.: Safety verification of hybrid systems by constraint propagation based abstraction refinement. ACM Transactions in Embedded Computing Systems 6(1) (2007)
54. Rizk, A., Batt, G., Fages, F., Soliman, S.: Continuous valuations of temporal logic specifications with applications to parameter optimization and robustness measures. Theor. Comput. Sci. 412(26), 2827–2839 (2011)
55. Ruohonen, K.: An effective Cauchy-Peano existence theorem for unique solutions. Internat. J. Found. Comput. Sci. 7(2), 151–160 (1996)
56. Spitters, B.: Constructive algebraic integration theory. Ann. Pure Appl. Logic 137(1-3), 380–390 (2006)
57. Taylor, P.: A lambda calculus for real analysis (2008), http://www.monad.me.uk/
58. Tran, D.X., Sato, D., Yochelis, A., Weiss, J.N., Garfinkel, A., Qu, Z.: Bifurcation and chaos in a model of cardiac early afterdepolarizations. Phys. Rev. Lett. 102(25), 258103 (2009)
59. Weihrauch, K.: Computable Analysis: An introduction. Texts in Theoretical Computer Science. Springer, Berlin (2000)
60. Zgliczynski, P., Kapela, T.A.: Lohner algorithm for perturbation of ODEs and differential inclusions. Discrete Contin. Dyn. Syst. Ser. B 11(2), 365–385 (2009)
61. Zgliczyński, P., Mischaikow, K.: Rigorous numerics for partial differential equations: The KuramotoSivashinsky equation. Found. Comput. Math. 1, 255–288 (2001)
62. Živanović, S., Collins, P.: Numerical solutions to noisy systems. In: Proc. 49th IEEE Conference on Decision and Control (2010)

Developing Quantitative Methods in Community Ecology: Predicting Species Abundances from Qualitative Web Interaction Data

Hugo Fort

Department of Physics, Faculty of Sciences, Universidad de la República,
Iguá 4225, Montevideo 11400, Uruguay
hugo@fisica.edu.uy

Abstract. Quantitative predictions of biodiversity of human-impacted ecological communities are crucial for their management. In the case of plant–pollinator mutualistic networks, despite the great progress in describing the interactions between plants and their pollinators, the capability of making quantitative predictions is still in its infancy. Furthermore, a general problem is the lack of measures or estimations of species abundances.

Here I propose a general method to estimate pollinator species abundances and their niche distribution from the available data, namely network interaction matrices. It works by transforming a plant–pollinator network into a competition model between pollinator species. Competition matrices were obtained from 'first principles' calculations, using qualitative interaction matrices compiled for a set including more than 40 plant–pollinator networks. This method is able to make accurate quantitative predictions for mutualistic networks spanning a broad geographic range. Specifically, the predicted biodiversity metrics for pollinators – species relative abundances, Shannon equitability and Gini–Simpson indices – agree quite well with those inferred from empirical counts of visits of pollinators to plants.

The importance of interspecific competition between pollinator species is a controversial and unresolved issue, considerable circumstantial evidence has accrued that competition between insects does occur, but a clear measure of its impact on their species abundances is still lacking. The present work contributed to fill this gap by quantifying the effect of competition between pollinators.

Particular applications could be to estimate the quantitative effects of removing a species from a community or to address the fate of populations of native organisms when foreign species are introduced to ecosystems far beyond their home range. This method also allows building a one-dimensional niche axis for pollinators in which clusters of generalists are separated by specialists thus rendering support to the theory of emergent neutrality.

Keywords: Quantitative ecology, species abundances, Lotka-Volterra competition.

F. Fages and C. Piazza (Eds.): FMMB 2014, LNBI 8738, pp. 23–35, 2014.
© Springer International Publishing Switzerland 2014

1   Introduction

Quantitative predictions of biodiversity of human-impacted ecological communities are crucial for their management. In the case of plant–pollinator mutualistic networks, despite the great progress in describing the interactions between plants and their pollinators [1], the capability of making quantitative predictions is still in its infancy [1,2]. To assess the impact of management decisions on communities after the loss or gains of species, it is necessary to go beyond the qualitative biodiversity analysis of general and abstract communities and be able to make quantitative predictions for real and specific communities found in nature. For example, understanding species abundances in mutualistic plant-pollinator webs, is currently limited mainly because a lack of 1) estimation of species abundances independent from recorded interactions between pollinators and plants [3] and 2) methodological tools to deal with the intrinsic complexity of mutualisms in general and pollination in particular [1]. A panel convened by the NSF in 2006 to discuss the "frontiers of ecology", and to make recommendations for research priority areas in population and community ecology, stated that ecology will become a more quantitative and predictive discipline if research is focused on the strength of interactions between species [4].

Here I propose a general method to estimate pollinator species abundances and their niche distribution from the available data, namely network interaction matrices. It works by transforming a plant–pollinator bipartite network into a unipartite competition model between pollinator species. A rationale for this approach is that a considerable amount of evidence suggests that competition for floral resources between pollinators is important and widespread [5-7] because the availability of partner plants represents a limiting resource [8]. The shared use of common resources results in interference and exploitative competition among pollinators that not only reduces foraging efficiency at feeding patches [9,10] but also pollen and nectar harvest of colonies [11,12]. Availability of either foods or nest sites is thus likely to limit population density of some pollinator species [13,14].

Our starting point is the Lotka-Volterra competition equations (without density-dependent terms regulating the growth of species) for the abundances N_a of the S_A animal species:

$$\frac{dN_a}{dt} = r_a N_a \left(1 - \sum_{a'=1}^{S_A} \alpha_{aa'} \frac{N_{a'}}{K_a} \right), \qquad a = 1, \cdots, S_A, \qquad (1)$$

where r_a and K_a denote the logistic coefficients, respectively the maximum growth rate and the carrying capacity, and $\alpha_{aa'}$ the competition coefficient between animal species a and a'. Since we are interested in the stationary behavior we took $r_a = 1$ for all the animal species. Additionally notice that in order to predict the relative abundances of pollinator species (RAPS), n_a, only the relative values of their carrying capacities, $k_a = K_a / K_{min}$, are required rather than the values of the carrying capacities themselves. So eq. (1) in terms of relative quantities becomes

$$\frac{dn_a}{dt} = n_a \left(1 - \sum_{a'=1}^{S_A} \alpha_{aa'} \frac{n_{a'}}{k_a} \right), \qquad a = 1, \cdots, S_A, \tag{2}$$

We will see that the two sets of parameters, $\{k_a\}$ and $\{\alpha_{aa'}\}$, can be simply estimated from the $S_A \times S_P$ empiric *qualitative interaction matrices* g_{ap} between S_A animal species and S_P plant species associated to each mutualistic network [15,16]. A given g_{ap} matrix has rows and columns corresponding to the animal and plant species of the system respectively and 1's and 0's in the row-column intersections represent interactions or their absence, respectively. The number of plant species with which the animal species a interacts, i.e. the number of 1's in a row of g_{ap}, is called its degree D_a [17]. The higher (lower) this index is, the more generalist (specialist) the pollinator species is.

In order to test the goodness of this method we need empiric data on species relative abundances to compare against the n_a. Unfortunately, there are no estimations of species abundances for pollinator species independent from recorded interactions between pollinators and plants [3]. The best available information for some mutualistic networks is comprised in *quantitative interaction matrices*, q_{ap}, which for each mutualistic network specifies the number of observed visits of each animal species a (row) to each plant species p (column). In these cases, for a given pollinator species a, one can estimate from q_{ap} its relative abundance, n_a, simply as proportional to the sum of the matrix elements over the corresponding row a i.e. all the recorded visits of this animal species to the different species of plants [7]. It turns out that the predicted RAPS, the Shannon equitability and the Simpson-Gini index agree quite well with those inferred from the q_{ap} matrices from a dataset including 13 mutualistic networks spanning a broad geographical range (see below). The good matching for these 13 networks, for which there is quantitative data on plant-pollinator interactions, suggests that the proposed approach could reliably predict biodiversity metrics for mutualistic networks for which these quantitative interaction matrices are not available.

2   Dataset and Parameter Estimation

2.1   Dataset

The data set of plant-pollinator interactions comprises a total of 13 mutualistic networks. It was compiled using the data from [15] plus data from [16]. For all these networks, spanning a broad geographic range from arctic to tropical and temperate, in addition to qualitative matrices g_{ap}, there are also quantitative interaction matrices, q_{ap}. Table 1 lists these networks, including their number of animal species, S_A, and the corresponding locations.

Table 1. The 13 analyzed plant-pollinator mutualistic networks

Number	Code name	S_A	Locality
1	BAHE	102	Central New Brunswick, Canada
2	BEZE	13	Pernambuco State, Brazil
3	DIHI	61	Hickling, Norfolk, UK
4	DISH	36	Shelfanger, Norfolk, UK
5	INPK	85	Snowy Mountains, Australia
6	KT90	679	Ashu, Kyoto, Japan
7	MEMM	79	Bristol, England
8	MOMA	18	Melville Island, Canada
9	MOTT	44	North Carolina, USA
10	OLLE	56	KwaZulu-Natal region, South Africa
11	SCHM	33	Brownfield, Illinois, USA
12	SMAL	34	Ottawa, Canada
13	VASI	30	Llao Llao, Argentina

2.2   Estimation of Competition Coefficients

Competition coefficients $\{\alpha_{aa'}\}$ were obtained from 'first principles' calculations, using the g_{ap} matrices compiled for a dataset of more than 40 plant–pollinator networks compiled from [15,15]. The recipe is straightforward and based on MacArthur and Levins idea [18] that the interspecific competition coefficient $\alpha_{aa'}$ between animal species a and a' is proportional to the number of shared resources (plants), or their resource overlap, as schematically shown in Fig. 1.

The resource overlap between species a and a' can be quantified by the Jaccard similarity index [19] $J_{aa'}$, which can be easily obtained from g_{ap} as:

$$J_{aa'} = \frac{P_{aa'}^{11}}{P_{aa'}^{01} + P_{aa'}^{10} + P_{aa'}^{11}}, \tag{3}$$

Fig. 1. The **intensity of competition between pollinator species depends on the number of plants they share.** The resource utilization for each pollinator species is represented by the set of species of plants it feeds on inside a box. In this example, taken from MEMM network, pollinator species *Anglais urticae* feeds on three species of plants (red filled box), i.e. it has $D = 3$, while pollinator species *B. muscorum* feeds on two species of plants (blue dashed box), i.e. it has $D = 2$. Therefore the overlap between these two pollinator species is one plant species.

where $P_{aa'}^{11}$ represents the number of species of plants visited by both species of animals and $P_{aa'}^{10}$ ($P_{aa'}^{01}$) the number of species of plants which are visited only by a (a'). Notice that if the two pollinator species share all the plants, i.e. $P_{aa'}^{10} = 0 = P_{aa'}^{01}$, then $J_{aa'} = 1$. This situation of full overlap in general occurs between maximum specialists with $D = 1$. Therefore,

$$\alpha_{aa'} = N_{aa'} J_{aa'}, \tag{4}$$

where the factors $N_{aa'}$ are equal to 1 for $a = a'$ in such a way that intraspecific competition coefficients $\alpha_{aa} = J_{aa}$ become 1. The non diagonal factors $N_{aa'}$ were chosen as random numbers from a uniform distribution and serve two different purposes. First, to include the fact that interspecific competition between pollinators is in general highly asymmetrical [20]. This asymmetry is known to be particularly important in the case of interference competition i.e. when each species harm the other directly [21]. Nevertheless, here we did not distinguish between interference competition and exploitative competition (when each species alter the abundance of some shared resource to the other). Second, to take into account different mechanisms that tend to weaken interspecific competition between animals, such as facilitation between them [20,22] or resource partitioning [22,23] these factors $N_{aa'}$ must be smaller than 1.

2.3   Estimation of Relative Carrying Capacities

The $\{k_a\}$ in turn were obtained from an empiric fact: the correlation between the observed empiric relative abundances and degree were large and significant [24] (see Fig. 2).. This same correlation was found by Suweis *et al.* [25]. Indeed MacArthur [26] and

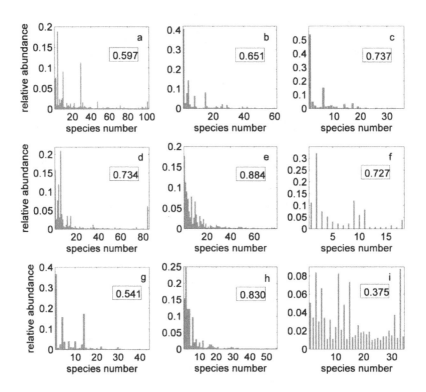

Fig. 2. Empiric RAPS for nine of the 13 plant-pollinator networks analyzed: BAHE (a), DIHI (b), DISH (c), INPK (d), MEMM (e), MOMA (f), MOTT (g), OLLE (h) & SMAL (i). Insect species are numbered according to decreasing degree D_a. The numbers in boxes correspond to the correlation between n_a and D_a, which is strong i.e. > 0.6 in most cases and in all the cases is significant (pValue < 0.01 for (a)-(h) and pValue < 0.05 for (i)).

later Brown [27] suggested that species with a broad ecological niche would, as a consequence, be able to reach higher abundances than species with a narrow niche breadth. Such a positive relationship between niche width or generalization level in resource use and relative abundance of individual species suggested that we can take k_a just identical to the degree D_a, an assumption that turned out to work pretty well (see below).

3   Results

3.1   Simulations

For each network 200 simulations were run for computing the mean abundances of each pollinator species and their standard deviations. There were two sources of stochasticity: Firstly, in the initial conditions, each simulation starts from a random abundance for each animal species. In each simulation a rapid convergence (in less than 100 temporal steps) to the equilibrium abundances was observed. Secondly, the non diagonal random normalizing factors $N_{aa'}$ were drawn at the beginning of each simulation.

3.2   Comparison between Empirical and Theoretical RAPS

The empirical RAPS were obtained simply from q_{ap} as

$$n_a^{emp} = \frac{\sum_{p=1}^{S_P} q_{ap}}{\sum_{a=1}^{S_A} \sum_{p=1}^{S_P} q_{ap}} \qquad a = 1, \cdots, S_A. \tag{5}$$

To measure the agreement between empirical and theoretical RAPS we used two indices between 0 and 1 (1 indicating perfect agreement between predicted and observed data and 0 indicating complete disagreement.).
First, the Willmott index of agreement $d_{emp\text{-}theo}$ (1981):

$$d_{emp\text{-}theo} = \frac{\sum_{a=1}^{S_A} (n_a^{theo} - n_a^{emp})^2}{\sum_{a=1}^{S_A} \left(\left| n_a^{theo} - \left\langle n_a^{theo} \right\rangle \right| + \left| n_a^{emp} - \left\langle n_a^{emp} \right\rangle \right| \right)^2}. \tag{6}$$

From (6) it can be derived the expected value of d for two independent random variables, $d_{random} = 3/7 \cong 0.429$.
Second, the Pearson correlation coefficient $r_{emp\text{-}theo}$.

3.3   Predicted vs. Empirical Biodiversity Metrics

The predicted RAPS were in general in quite good agreement with the empirical ones (Fig.2 and Table 1). For the 13 networks $d_{emp\text{-}theo} > 0.5$ and in general well above 0.5

(see Table 1). Other biodiversity metrics that can be built from the $\{n_a\}$ are the Shannon equitability and the Simpson-Gini index defined, respectively, by

$$H = \sum_a^{S_A} n_a \ln n_a / S_A \text{ and } SG = 1 - \sum_a^{S_A} n_a^2.$$

We can see that the best matching between theoretical predictions and empirical data for all the biodiversity metrics –RAPS, H, SG– occurs for the MEMM network (Table 2), while the worst matching occurs for the SMAL network.

Table 2. Empirical and theoretical biodiversity metrics for plant-pollinator networks. For all the networks, except for the SMAL network, p-value < 0.01 and the correlation was significant.

Network code [(*)]	Richness S_A [(*)]	Agreement emp. & theo. RAPS		Shannon's equitability		Simpson-Gini index	
		$d_{emp\text{-}theo}$	$r_{emp\text{-}theo}$	H_{emp}	H_{theo}	SG_{emp}	SG_{theo}
BAHE	102	0.74	0.59	0.74	0.69	0.93	0.93
BEZE	13	0.98	0.97	0.79	0.75	0.82	0.82
DIHI	61	0.77	0.67	0.56	0.66	0.80	0.90
DISH	36	0.78	0.78	0.53	0.76	0.68	0.90
INPK	85	0.81	0.72	0.66	0.77	0.90	0.95
KT90	679	0.78	0.70	0.84	0.67	0.99	0.97
MEMM	79	0.96	0.91	0.72	0.71	0.93	0.93
MOMA	18	0.85	0.74	0.79	0.83	0.85	0.87
MOTT	44	0.63	0.54	0.59	0.80	0.80	0.94
OLLE	56	0.89	0.88	0.65	0.77	0.87	0.94
SCHE	32	0.83	0.72	0.73	0.78	0.87	0.91
SMAL	34	0.51	0.28	0.93	0.89	0.95	0.94
VASI	30	0.81	0.79	0.58	0.80	0.78	0.91

The highest concordance between the empirical and theoretical relative abundances occurred either for highly specialized pollinator species (low D_a), unless two exceptions. These two exceptions for which the method fails to reproduce the abundance of specialist insect species with relative abundance above 5 %, occur one for the INPK network (Fig. 2-d) and one for the SMAL network (Fig. 2-i). How to fix these significant underestimations of specialist species is treated in the Discussion section.

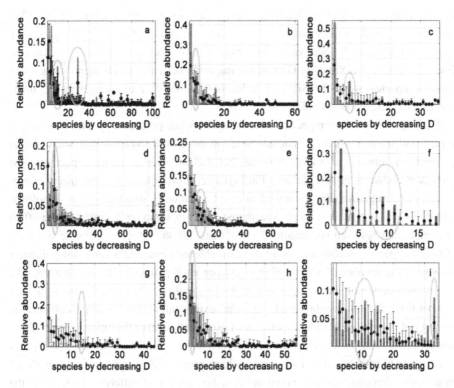

Fig. 3. Predicted and observed RAPS for nine of the 13 plant-pollinator networks analyzed. Empiric (gray bars) and theoretical (black with errors = 1 std for 100 simulations) RAPS for: BAHE (a), DIHI (b), DISH (c), INPK (d), MEMM (e), MOMA (f), MOTT (g), OLLE (h) & SMAL. Insect species are numbered according to decreasing degree D_a. See text in Discussion for the ovals.

4   Discussion

In order to evaluate the impact of human activities on ecological communities, that lead to the loss or gains of species, and to decide on the most effective actions for nature conservation, we need to develop quantitative prediction tools. In the specific case of plant-pollinator communities predictive capacities are limited to qualitative assessments [1].

I presented a simple method to make quantitative predictions about the animal relative abundances from the qualitative information of their animal-plant mutualistic networks summarized in the adjacency matrices g_{ap}. Even though the theoretical parameters were estimated from such coarse (binary) data, the predicted RAPS are in general in quite good agreement with the empirical ones –estimated from the quantitative interaction matrices q_{ap}– for mutualistic networks spanning a broad geographic range. The comparison between theoretical and empirical biodiversity indices shows a better agreement for the Simpson-Gini than for the Shannon equitability. This can be easily understood from the fact that the theoretical RAPS are averages over samples of 200 simulations whose rare species have larger coefficients of variation. Therefore, since the Simpson-Gini index responds basically to changes in the most common species, while the Shannon index is more sensible to changes in the abundance of the rarest species [28].

The importance of interspecific competition between pollinator species is a controversial and unresolved issue, considerable circumstantial evidence has accrued that competition between insects does occur, but a clear measure of its impact on their species abundances is still lacking. Furthermore, little is known about direct consequences of competition for the abundance and community structure of insect pollinators [8,29,30]. The present work contributed to fill this gap by quantifying the effect of competition between pollinators. That is, competition matrices were obtained from "first principles" calculations, using the available empirical data for plant-pollinator networks: the number of shared plants between each pair of pollinator species. The higher the fraction of shared plant species by two pollinator species, the stronger their competition. Yet one might wonder to what extent taking into account competition among pollinators really increases the fit of estimated relative densities compared to a neutral model with no interspecific interactions and k_a identical to the degree D_a plus some sort of random noise. However, without noise, such a model would predict a monotonic abundance-D functional relationship, which is different from what the empirical data show. That is: a) species with low D are frequently more abundant than other with high D, and b) species with equal D sometimes exhibit important differences in their abundances. It turns out that the interspecific competition interactions can explain both a) and b). In fig. 2 the ovals point entire sectors of the RAPS departing from the neutral monotonic behavior that can be reproduced by the proposed competition model. The addition of noise to the neutral model would hardly reproduce these empirical observations.

The proposed model can be improved by incorporating into it additional empirical available information on species abundances. For instance this model fails to reproduce few abnormal cases of abundant specialist insect species (e.g. for the SMAL network). A simple explanation for this mismatch is that the floral host, *Nemopanthus mucronata*, of this specialist, *Dilophus caurinus* [15], be highly abundant (unfortunately, as far as I know, there are no data on the abundance of plants). So a simple way to improve the model predictions in these cases is to incorporate either plausible or known empirical facts yielding a higher level model. For instance, if the floral resource of a specialist is known to be highly abundant, for this species it is sensible to go beyond the simple rule of taking the carrying capacity as proportional to the species degree and rather take a larger value for its carrying capacity. Fig.6 shows how the predictions of the model for the SMAL network improve by changing the carrying capacity of *Dilophus caurinus*, from $k_{Dc} = D_{Dc} = 1$ to $k_{Dc} = 5$ (since no data on this plant abundance were available the value of k_{Dc} was determined by trial and error to achieve the best agreement with the estimated empirical relative abundance for *Dilophus caurinus*). Notice that, besides the dramatic improvement for the predicted abundance of this particular species, the matching between theoretical and empirical relative abundances of all the other species also improves in some degree.

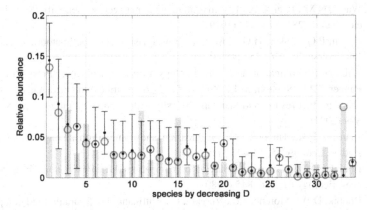

Fig. 4. Improvements in the predicted RAPS when changing the carrying capacity of one species. The same RAPS from Fig. 2-i: empiric (gray bars), theoretical predictions without change (black dots with errors = 1 std for 100 simulations); plus the predicted RAPS when changing the relative carrying capacity of *Dilophus caurinus* from $k_m = 1$ to $k_m = 5$ (large gray circles).

Practical applications of this method to community management could be to estimate quantitative effects of removing a species from a community or to address the fate of populations of native organisms when foreign species are introduced to ecosystems far beyond their home range.

Last but not least, this approach also allows the construction of the distribution of pollinator niches by grouping together all the species with large overlaps. An interesting pattern emerges: clusters of generalists are separated by specialists [24]. The niche framework helps to devise each specie's place and role in the community to which it belong as well as it provides a better understanding of species coexistence, new insights and interpretations about ecological patterns and processes in ecology and how fluctuating environments might regulate population dynamics and species interactions.

Acknowledgements. Work supported in part from PEDECIBA (Uruguay) and ANII (SNI Uruguay).

References

[1] Bascompte, J., Jordano, P.: Mutualistic Networks. Princeton University Press (2013)

[2] Morin, P.J.: Community Ecology, 2nd edn. John Wiley & Sons (2011)

[3] Blüthgen, N.: Why network analysis is often disconnected from community ecology: a critique and an ecologist's guide. Basic and Applied Ecology 11(3), 185–195 (2010)

[4] NSF, panel on "frontiers of ecology" (2006), http://www.nsf.gov/funding/pgm_summ.jsp?pims_id=12823&org=DEB&from=home

[5] Waser, N.M., Real, L.A.: Effective mutualism between sequentially flowering plant species. Nature 281, 670–672 (1979)

[6] Kevan, P.G., Baker, H.G.: Insects as flower visitors and pollinators. Annual Review of Entomology 28, 407–453 (1983)

[7] Willmer, P.: Pollination and Floral Ecology. Princeton University Press, Princeton (2011)

[8] Palmer, T.M., Stanton, M.L., Young, T.P.: Competition and coexistence: ex-ploring mechanisms that restrict and maintain diversity within mutualist guilds. Am. Nat. 162, S63–S79 (2003)

[9] Johnson, L.K., Hubbell, S.P.: Aggression and competition among stingless bees: field studies. Ecology 55, 120–127 (1974)

[10] Roubik, D.W.: Foraging behavior of competing Africanized honeybees and stingless bees. Ecology 61, 836–845 (1980)

[11] Roubik, D.W., Moreno, J.E., Vergara, C., Wittmann, D.: Sporadic food competition with the African honeybee: projected impact on Neotropical social bees. J. Trop. Ecol. 2, 97–111 (1986)

[12] Wilms, W., Wiechers, B.: Floral resource partitioning between native Melipona bees and introduced Africanized honeybee in the Brazilian Atlantic rain forest. Apidologie 28, 339–355 (1997)

[13] Hubbell, S.P., Johnson, L.K.: Competition and nest spacing in a tropical stingless bee community. Ecology 58, 949–963 (1977)

[14] Inoue, T., Nakamura, K., Salmah, S., Abbas, I.: Population dynamics of animals in unpredictably-changing tropical environments. J. Biosciences 18, 425–455 (1993)

[15] Rezende, E., Lavabre, J.E., Guimarães, P.R., Jordano, P., Bascompte, J.: Non-random coextinctions in phylogenetically structured mutualistic networks. Nature 448, 925–928 (2007), Supplementary Data and Methods: http://www.nature.com/nature/journal/v448/n7156/suppinfo/nature05956.html

[16] NCEAS 2014 (2014), http://www.nceas.ucsb.edu/interactionweb/resources.html

[17] Jordano, P., Bascompte, J., Olesen, J.M.: Invariant properties in coevolutio-nary networks of plant–animal interactions. Ecol. Lett. 6, 69–81 (2003)

[18] MacArthur, R.H., Levins, R.: The limiting similarity, convergence and diver-gence of coexisting species. Am. Nat. 101, 377–385 (1967)

[19] Jaccard, P.: Étude comparative de la distribution florale dans une portion des Alpes et des Jura. Bulletin de la Société Vaudoise des Sciences Naturelles 37, 547–579 (1901)

[20] Kaplan, I., Denno, R.F.: Interspecific interactions in phytophagous insects revisited: a quantitative assessment of competition theory. Ecology Letters 10, 977–994 (2007)

[21] Keddy, P.A.: Competition, 2nd edn., pp. 333–404. Kluwer Academic Publishers, Dordrecht (2001)

[22] Denno, R.F., Kaplan, I.: Plant mediated interactions in herbivorous insects: mechanisms, symmetry and challenging the paradigms of competition past. In: Ohgushi, T., Craig, T.P., Price, P.W. (eds.) Ecological Communities: Plant Mediation in Indirect Interaction Webs, pp. 19–50. Cambridge University Press, London (2007)

[23] Lande, R.: Statistics and partitioning of species diversity, and similarity among multiple communities. Oikos 76, 5–13 (1996)

[24] Fort, H.: Quantitative Predictions of Pollinators' Abundances from Qualitative Data on their Interactions with Plants and Evidences of Emergent Neutrality. Oikos (2014), doi:10.1111/oik.01539

[25] Suweis, S., Simini, F., Banavar, J.R., Maritan, A.: Emergence of structural and dynamical properties of ecological mutualistic networks. Nature 500, 449–452 (2013)

[26] MacArthur, R.H.: Geographical ecology. Harper and Row, New York (1972)

[27] Brown, J.H.: On the relationship between abundance and distribution of species. Am. Nat. 122, 295–299 (1984)

[28] Peet, R.K.: The measurement of species diversity. Annual Review of Ecology and Systematics 5, 285–307 (1974)

[29] Steffan-Dewenter, I., Tscharntke, T.: Resource overlap and possible competition between honeybees and wild bees in central Europe. Oecologia 122, 288–296 (2000)

[30] Goulson, D.: Effects of introduced bees on native ecosystems. Annual Review of Ecology. Evolution and Systematics 34, 1–26 (2003)

Understanding How Biodiversity Is Distributed in Space and Time

Hélène Morlon

Institut de Biologie de l'Ecole Normale Supérieure, UMR 8197 CNRS
46 rue d'Ulm, 75005 Paris, France
morlon@biologie.ens.fr
http://www.biologie.ens.fr/phyloeco/index.html

Abstract. Attention to biodiversity issues has been growing in the recent years. Despite the urgency of the problem, the development of a general theory of biodiversity is still underway. How can we develop such a theory? Two main approaches have dominated the field: the first approach has emphasized ecological controls of biodiversity, and has sought to explain static biodiversity patterns, often referred to as macroecological patterns; the second approach has emphasized historical controls of biodiversity, and has sought to explain more dynamic, evolutionary patterns of biodiversity, often referred to as macroevolutionary patterns. I will quickly summarize these approaches, focusing on the analytical and computational methods that have been used. Then, I will discuss how we can hope to integrate these two approaches to obtain a theory of biodiversity that better accounts for both historical and ecological factors.

Keywords: biodiversity theory, macroecology, macroevolution.

Ecology has started with a Linnaean period of discovering, naming, and counting species on Earth. From this descriptive effort, universal patterns have emerged, such as the way species richness increases with area, the way species abundances are distributed within communities, the way community composition changes from place to place, or the way species richness varies along latitudinal, altitudinal or environmental gradients [1]. Understanding these universal macroecological patterns has been the focus of early theoretical work, which has therefore aimed to predict static diversity patterns. To illustrate the type of models and the analytical and computational methods that ecologists resort to predict these macroecological patterns, I will focus on one of the theories that has received most attention from theoreticians: the neutral theory of biodiversity (Fig. 1, [2]).

At the same time that ecologists documented biodiversity patterns for existing species across the surface of the Earth, paleobiologists documented temporal patterns in the fossil record, such as how the number of species, or genera, varied through time. Understanding these temporal dynamics has spurred the development of another type of models, so-called birth-death models of cladogenesis. Later, these macroevolutionary models have been used far beyond the field of paleobiology, becoming central to most macroevolutionary studies [4]. In particular, they have successfully been applied to analyse long-term diversity dynamics

F. Fages and C. Piazza (Eds.): FMMB 2014, LNBI 8738, pp. 36–39, 2014.

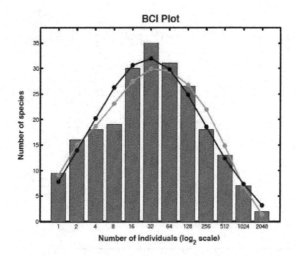

Fig. 1. An example of macroecological data used to test biodiversity theories. The red bars represent observed numbers of tree species binned into abundance categories in a 50-hectare forest plot in Barro Colorado Island. This macroecological pattern is well known as the species abundance distribution. The black and green curves represent the best fit of two different models. The neutral theory of biodiversity can be fitted to such data by maximum likelihood. From [3].

using molecular phylogenetic data. These models are now the basis of most phylogenetic methods used to understand why some groups of organisms are more species rich than others, why some geographic regions are more species rich than others, and what controls temporal variations in species richness [5]. I will detail the analytical and computational methods used to fit these macroevolutionary models to empirical data, focusing on work from my research group (Fig. 2, [6–10]).

While macroecological and macroevolutionary models have been developed separately, with little connections between the two fields, there is a need for an integration of these two types of models. Indeed, in their current form, macroecological models are not especially useful for understanding why species richness varies across taxonomic groups and geographic regions, while macroevolutionary models are not especially useful for understanding the distribution and abundance of species among communities. I will present news results in this direction, and discuss ways forward for a better integration of macroecological and macroevolutionary models.

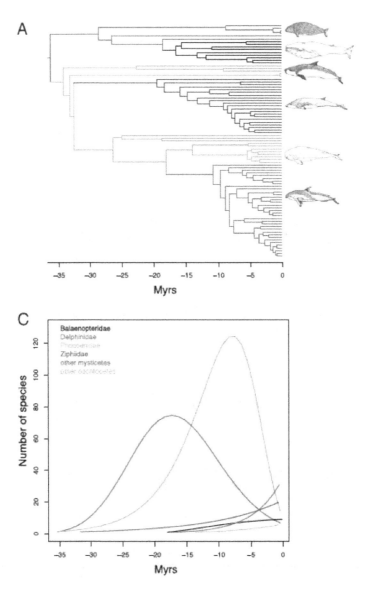

Fig. 2. An example of macroevolutionary data used to test biodiversity theories. The upper panel represents the phylogeny (dated evolutionary tree) of cetaceans (whales, dolphins and porpoises). The lower panel represents the dynamics of diversity (species richness through time) that can be inferred from this phylogenetic data, using birth-death models of cladogenesis. From [7].

References

1. Rosenzweig, M.L.: Species diversity in space and time. Cambridge University Press (1995)
2. Hubbell, S.P.: The unified neutral theory of biodiversity and biogeography. Princeton University Press (2001)
3. Volkov, I., Banavar, J.R., Hubbell, S.P., Maritan, A.: Neutral theory and relative species abundance in ecology. Nature 424, 1035–1037 (2003)
4. Nee, S.: Birth-death models in macroevolution. Ann. Rev. Ecol. Evol. Syst., 1–17 (2006)
5. Morlon, H.: Phylogenetic approaches for studying diversification. Ecol. Lett., 508–525 (2014)
6. Morlon, H., Potts, M.D., Plotkin, J.B.: Inferring the dynamics of diversification: a coalescent approach. PLoS B., e1000493 (2010)
7. Morlon, H., Parsons, T.L., Plotkin, J.B.: Reconciling molecular phylogenies with the fossil record. Proc. Nat. Acad. Sci. USA, 16327–16332 (2011)
8. Condamine, F.L., Rolland, J., Morlon, H.: Macroevolutionary perspectives to environmental change. Ecol. Lett., 72–85 (2013)
9. Rolland, J., Condamine, F.L., Jiguet, F., Morlon, H.: Faster speciation and reduced extinction in the tropics contribute to the mammalian latitudinal diversity gradient. PLoS Biology, e1001775 (2014a)
10. Rolland, J., Jiguet, F., Jonsson, K.A., Condamine, F.L., Morlon, H.: Settling down of seasonal migrants promotes bird diversification. Proc. Royal Soc. B, 20140473 (2014b)

Computing Longevity: Insights from Controls

Pietro Lió

Computer Laboratory, University of Cambridge, UK
{pietro.lio}@cl.cam.ac.uk

Abstract. There is a growing perception that medical treatment could be effective against aging although not as one intensive short time medication as we do with infections. It will require a precise, personalised knowledge of the genes and pathways that are perturbed during the progression to aging. Environmental factors, parental longevity and childhood are important predictors of exceptional longevity. Here we analyse molecular data (gene expression) from "healthy" controls of different age from several studies and we identify perturbations in key pathways affecting the susceptibility to several diseases. This work is exploratory and provide a useful test on existing data and methods for future studies.

1 Introduction

Despite the fact that all our atoms are continuously recycled and no remains in our body for more than two decades, for humans, living to 100 years is a rare event and is linked to aging. Aging occurs in every multicellular animal that reaches a fixed size at reproductive maturity. In human, the variance in life expectation depends on the longevity of parents and to some childhood and midlife epigenetic and metabolic characteristics. Aging could be measured with various physiological, cognitive, and behavioral tests, for instance the decrease in muscle strength and size, in skin thickness and in the ability to learn. With aging progression there is an increase risk for many diseases, including Alzheimer, diabetes, cardiovascular disorders, and cancer. In addition to the progressive failure of the body's repair mechanisms during aging, there is the spreading of multi morbidities and comorbidities which have often shared molecular pathways [2] [1]. The systemic inflammation (inflammaging) and the arising of qualitative and quantitative changes in the three types of DNA (nuclear, mitochondrial and gut bacterial) have shown to be important factors in the aging process [3]. The knowledge of the pathways disrupted during aging could provide clue on how to use gene and stem cell therapy and new mitochondria or tissue bioengineering to increase the human longevity [4]. Epidemiological and meta-analysis of molecular data across ages provide important clue for longevity [5]. Here we consider gene expression data of 'controls', i.e. relatively healthy persons used for comparison in many disease studies. We analyse the disruption of genes and pathways showing positive or negative trends across ages. Given the complexity of the topic, this study is exploratory.

F. Fages and C. Piazza (Eds.): FMMB 2014, LNBI 8738, pp. 40–46, 2014.

2   Methods and Datasets

2.1   Data Used

We used data obtained from energy demanding tissues such as muscle and brain, and also from blood cells. Males and females groups were treated separately. We downloaded the following several datasets which include a good number of controls:

- Muscle - GEO accessions: GDS472, GSD473 [6]. Both sets feature two female age groups: 20-29yrs and 65-71yrs, with 7 and 8 individual samples in each group respectively; male, GEO accessions: GDS287, GDS288 [7]. Both sets feature two age groups: 21-27yrs and 67-75yrs, with 7 and 8 individual samples in each group respectively. Other datasets: GSE5086, GSE38718 (see also [8]).
- Brain - frontal cortex expression profiles at various ages, GEO accession: GDS707 [9]. It consists of 30 samples of healty individuals, 18 males and 12 females between 26-106 years. The male cohort had information over age groups 20-39, 40-59, 60-79 and 80-99 yrs. The female cohort had information over age groups 20-39, 40-59, 60-79, 80-99 and 106 yrs. The male cohort on the other hand had at least four replicates in each age group. We analysed the North American Brain Expression Consortium and UK Human Brain Expression Database with 991 healthy human brain samples (455 taken from the frontal cortex and 456 taken from the cerebellum). The data set can be found at http://www.ebi.ac.uk/arrayexpress/experiments/E-GEOD-36192/; GSE46706 [10]; other datasets are GSE11504, GSE36192,GSE55457, GSE9348.
- Blood and immune system - Data used in [11] and [12] were downloaded from the Gene Expression Omnibus (GEO) data repository; we used also type II Diabetes dataset (GSE38642, [13]). This dataset contains information about 63 individuals, 54 of which are controls (non-diabetic) and contains information on 23 control females and 31 control males with age range from 26 to 75 years.

2.2   Methods

Bioconductor [14] and Limma [15] were used to identify differentially expressed genes. The data was loaded and manipulated using GEOquery [16]. R was employed to derive heatmap and clustering analysis. Gene ontology information was available within the pheno-data of the expression set object. The gene IDs can also be queried on enrichment services such as DAVID (available at http://david.abcc.ncifcrf.gov/). The list of differentially expressed genes was also fed into KOBAS Web service [19]. This service can analyze enrichment of genes in disease databases such as KEGG DISEASE and OMIM. Gene ontology information may provide clues as to what comorbidities these genes are associated with. There are also many disease-gene association databases which could be queried with the gene list.

For all the datasets we have compared gene expression profiles between samples from different age categories and identified sets of genes whose expression is correlated with age group. Those genes consistently identified as over or under-expressed with age in most datasets will be taken as the final list of genes with age associated expression. Raw intensities must be normalised and probe sets summarised into gene-level intensities. For each gene, the log2 transformed intensities are linearly regressed and regression parameters estimated by least squares regression. The following linear regression model without interaction was used in order to obtain differential expressed genes between young and aged samples: $expression \approx \beta_0 + \beta_{sex}A_{sex} + \beta_{age}A_{age}$. Here, A_{sex} is a categorical variable which takes 1 for male and 0 for female, and A_{age} is the control's age. Then, 0 is the average gene expression, β_{sex} is the additional gene expression in male and β_{age} is the change of gene expression over age. For each regression, a two-tailed F test tests whether the gradient is different from zero. Multiple test correction is applied and the result is a list of genes that are either over or under expressed with aging [17]. When this method is applied to many datasets, the resulting set of lists of differentially expressed genes must be refined to find those genes differentially expressed across all (or the majority of) datasets. For each gene, there is a count of its identification as differentially expressed. This list of genes with age associated expression can then be manually validated, by procedures such as quantitative-PCR. Gene ontology enrichment and known gene-disease associations can be queried to assess comorbidities associated with the differentially expressed genes and associated with aging [18].

3   Results

The analysis of controls in multiple datasets shows that age and tissue type have a much larger effect on global gene expression profile in these samples than does the sex of the individual from which the sample was taken. In figure 1 we show the relationships of genes found in this study (only p-value lower than 0.001). As an example, we found a group of genes with statistically meaningful age dependent changes across the majority of datasets is involved in several "distributed" pathways (see Table 1).

The age-specific variations in the small number of genes above listed could influence the following pathways: PI3K signaling events mediated by Akt, S1P1 pathway, EGFR-dependent Endothelin signaling, Syndecan-1-mediated signaling, Glypican 1 network, Alpha9 beta1 integrin signaling, ErbB1 downstream signaling, Thrombin/protease-activated receptor, PDGF receptor signaling, LKB1 signaling, Arf6 signaling, mTOR signaling pathway, VEGF and VEGFR signaling, Alpha6Beta4 Integrin, Urokinase-type plasminogen activator and uPAR-mediated signaling, GMCSF-mediated signaling, Plasma membrane estrogen receptor signaling, IL5-mediated signaling, Hepatocyte Growth Factor Receptor signaling, Signaling events mediated by focal adhesion kinase, IGF1 pathway,

Table 1.

RHOQ	Ras homolog family member Q
ARAP2	ArfGAP with RhoGAP domain, ankyrin repeat and PH domain 2
MDM2	Mdm2, p53 E3 ubiquitin protein ligase homolog (mouse)
VLDLR	very low density lipoprotein receptor
CD79B	CD79b molecule, immunoglobulin-associated beta
EPS8	epidermal growth factor receptor pathway substrate 8
YWHAQ	tyrosine 3-monooxygenase/tryptophan 5-monooxygenase activation protein
FKBP5	FK506 binding protein 5
BCL2	B-cell CLL/lymphoma 2
CDKN1A	cyclin-dependent kinase inhibitor 1A (p21, Cip1)
ITGB4	integrin, beta 4
EIF4E	eukaryotic translation initiation factor 4E
DDIT4	DNA-damage-inducible transcript 4

PDGFR-beta signaling, Arf6 trafficking, Nectin adhesion pathway, IL3-mediated signaling, Insulin Pathway, IFN-gamma pathway, Sphingosine 1-phosphate pathway, Proteoglycan syndecan-mediated signaling, TRAIL signaling pathway, Endothelins. Other genes of Figure 1, among which BCL2,ccr6,cxcr5, AQP1, methyltransferase like 7A,IGLL1, add the influence to the following pathways: Beta1 integrin cell surface interactions, CDC42 signaling, glucocorticoid receptor signaling, p63 transcription factor network, citric acid cycle and respiratory electron transport, C-MYC pathway, p53 effectors, validated targets of C-MYC transcriptional repression, AP-1 transcription factor network, p53-Dependent G1/S DNA damage checkpoint, p53-dependent G1 DNA Damage Response, PIP3 activates AKT signaling,IL1-mediated signaling, insulin-mediated glucose transport, TGF-beta receptor signaling, Potassium Channels, BMP receptor signaling, C-MYB transcription factor network. For sake of space we will not discuss here each gene p value value and role; it is noteworthy that the AQP1 has a role as a gated ion channel; in astrocytomas, aquaporin 1 is expressed in microvessel endothelia and is involved in oedema., Data suggest that variations found in plasma osmolarity during hemodialysis may induce aquaporin 1 expression on the membrane of intact red blood cells. The immunoglobulin lambda-like polypeptide 1 (IGLL1) is critical for B-cell development. It is a preB cell receptor, found on the surface of proB and preB cells. It is involved in transduction of signals for cellular proliferation and differentiation from the proB cell to the preB cell. It is involved in light chain Ig rearrangements and allelic exclusion at the heavy chain Ig locus. Aging is accompanied by a decline in B lymphopoiesis in the bone marrow and accumulation in the periphery of B cells. One systemic aspect of comorbidity is inflammation that involves hundreds of molecules, and NFkβ is at the centre of that regulatory map. A recent work has shown that the activation of the NFkβ pathway in the hypothalamus of mice significantly accelerated the development of systemic aging that shortened the lifespan. The block of the NFkβ pathway in the hypothalamus of mouse brains slowed aging and increased median longevity by about 20 percent, compared to controls.

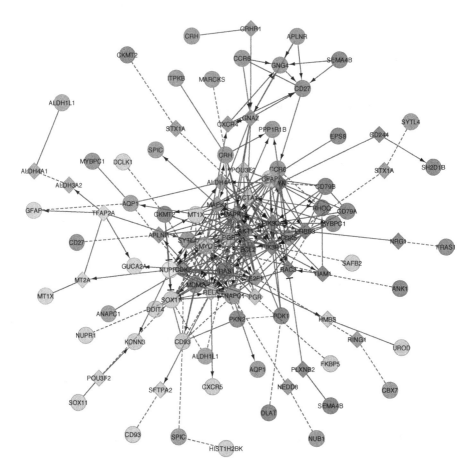

Fig. 1. Relationships between the genes found in work: ACOT7, AHBAK, ALDH1L1, ANAPC1, ANK1, AOAH, APLNR, AQP1, ARPP19, BCL2, C14orf159, C1orf162, CBX7, CCNY, CCR6, CD27, CD79A, CD79B, CD93, Cdc42, CDKN1A, CDR2, CKMT2, CRH, CTHRC1, CTIF, CX0rf57, CXCR5, DCLK1, DDIT4, DDIT4L, DLAT, DPP8, EMCN, EPS8, FAM102A, FKBP5, FRAS1, GFAP, GMPR, GNG4, GOLIM4, Grap, GRP, GUCA2A,H3F3B,HIST1H2BK, ITGB4,ITPKB, KAZALD1,KCNN3, KCTD20,LANCL2, LOC113230, LOC284023, MACROD1, MARCKS, MDM2,MS4A7, MT1A, MT1X,MYBPC1, MYC, NNT, NUB1, NUPR1, p21, PAK1, PKN2,PLEC, PPP1R1B, Pragmin, RHBDL3,RHOBTB3, RHOQ, SAFB2, SEMA4B,SERPINA3, SH2D1B, SLC14A1, SLC7A2, SOX11, SPATA2L, SPIC, STOM, SYTL4, TNS1, TPP1, UROD, VLDLR, WDR86,WIPI1, ZBTB16

4   Discussion

The sets of genes appear to influence a large number of pathways. This pervasiveness represents a sort of avalanche effect in aging, i.e. the large wired structure of the intracellular and intercellular signaling in our body, under aging produce large amount of comorbidities. It is noteworthy the decrease in disease connectivity shown by many centenarians with respect to the high wiring of people at earlier age (Franceschi Claudio personal communication). The pathway space is also fluid in the sense that other genes could replace the ones outlined by this study as major player in the pathway alteration. The increase of the number of centenarians and the slow or negligible in the number of supercentenarians may highlight other pathway factors may become predominant at later ages.

The pervasiveness of the pathway space corresponds to a pervasiveness in disease network space, i.e. comorbidities. From a molecular and cellular perspective, in aging there is a continuum of anatomical/pathological lesions and pathophysiological deficits that alters the homeostatic mechanisms for example, resilience to extremes of cold or heat, i.e. becoming less capable of absorbing shocks (Ben van Ommen, personal communication) and progresses from the pre-symptomatic stage beginning before birth through all the life, spreading from single disease to a variety of comorbidities [1]. Following this point of view, a good analogy of the aging process is the second law of thermodynamics due to the progressive decrease of our bodies ability to output entropy by cleaning metabolic debris and repairing the random damage occurring as a side-effect of metabolic activity.

Some brain tissues appear more revelatory than others. In particular a recent work found that the activating the NFkβ pathway in the hypothalamus caused declines in levels of gonadotropin-releasing hormone (GnRH), which is synthesized in the hypothalamus. Release of GnRH into the blood is usually associated with reproduction [20].

Future work should investigate the relationship between pathways disruption in different tissues and the arising of comorbidities.

References

1. Capobianco, E., Lió, P.: Comorbidity: a multidimensional approach. Trends in Molecular Medicine 19(9), 515–521 (2013)
2. Radner, H., Yoshida, K., Smolen, J.S., Solomon, D.H.: Multimorbidity and rheumatic conditions - enhancing the concept of comorbidity. Nature Reviews Endocrinology 10, 37–50 (2014)
3. Raule, N., Sevini, F., Li, S., Barbieri, A., Tallaro, F., Lomartire, L., Vianello, D., Montesanto, A., Moilanen, J.S., Bezrukov, V., Blanch, H., Hervonen, A., Christensen, K., Deiana, L., Gonos, E.S., Kirkwood, T.B., Kristensen, P., Leon, A., Pelicci, P.G., Poulain, M., Rea, I.M., Remacle, J., Robine, J.M., Schreiber, S., Sikora, E., Eline Slagboom, P., Spazzafumo, L., Antonietta Stazi, M., Toussaint, O., Vaupel, J.W., Rose, G., Majamaa, K., Perola, M., Johnson, T.E., Bolund, L., Yang, H., Passarino, G., Franceschi, C.: The co-occurrence of mtDNA mutations on different oxidative phosphorylation subunits, not detected by haplogroup analysis, affects human longevity and is population specific. Aging Cell 13(3), 401–407 (2013) (Epub. December 17, 2013), doi:10.1111/acel.12186
4. Garagnani, P., Pirazzini, C., Giuliani, C., Candela, M., Brigidi, P., Sevini, F., Luiselli, D., Bacalini, M.G., Salvioli, S., Capri, M., Monti, D., Mari, D., Collino, S., Delledonne, M., Descombes, P., Franceschi, C.: The three genetics (nuclear DNA, mitochondrial DNA, and gut microbiome) of longevity in humans considered as metaorganisms. Biomed Res. Int. 2014, 560340 (2014) (Epub. April 24, 2014), doi:10.1155/2014/560340

5. de Magalhes, J.P., Curado, J., Church, G.M.: Meta-analysis of age-related gene expression profiles identifies common signatures of aging. Bioinformatics 25(7), 875–881 (2009)
6. Welle, S., Brooks, A.I., Delehanty, J.M., Needler, N., et al.: Skeletal muscle gene expression profiles in 20-29 year old and 65-71 year old women. Exp. Gerontol. 39(3), 369–377 (2004), PMID: 15036396
7. Welle, S., Brooks, A.I., Delehanty, J.M., Needler, N., et al.: Gene expression profile of aging in human muscle. Physiol Genomics 14(2), 149–159 (2003), PMID: 12783983
8. Zahn, J.M., et al.: Transcriptional profiling of aging in human muscle reveals a common aging signature. PLoS Genetics 2(7), e115 (2006)
9. Lu, T., Pan, Y., Kao, S.Y., Li, C., et al.: Gene regulation and DNA damage in the ageing human brain. Nature 429(6994), 883–891 (2004), PMID: 15190254
10. Ryten, M., Trabzuni, D., Walker, R., et al.: Expression data generated from postmortem human brain tissue originating from neurologically and neuropathologically control individuals collected from: `http://www.ncbi.nlm.nih.gov/geo/query/acc.cgi?acc=GSE46706`
11. de Jong, S., Boks, M.P., Fuller, T.F., Strengman, E., et al.: A gene co-expression network in whole blood of schizophrenia patients is independent of antipsychotic-use and enriched for brain-expressed genes. PLoS One 7(6), e39498 (2012), PMID: 22761806
12. Linton, P.J., Dorshkind, K.: Age-related changes in lymphocyte development and function. Nat. Immunol. 5, 133–139 (2004)
13. Taneera, J., Lang, S., Sharma, A., Fadista, J., Zhou, Y., Ahlqvist, E., Jonsson, A., Lyssenko, V., Vikman, P., Hansson, O., Parikh, H., Korsgren, O., Soni, A., Krus, U., Zhang, E., Jing, X.J., Esguerra, J.L., Wollheim, C.B., Salehi, A., Rosengren, A., Renstrom, E., Groop, L.: A systems genetics approach identifies genes and pathways for type 2 diabetes in human islets. Cell Metabolism 16(1), 122–134 (2012)
14. Gentleman, R., Carey, V.J., Bates, D.M., Bolstad, B., Dettling, M., Dudoit, S., Ellis, B., Gautier, L., Ge, Y., et al.: Bioconductor: Open software development for computational biology and bioinformatics. Genome Biology 5, R80 (2004)
15. Smyth, G.K.: Limma: linear models for microarray data. In: Gentleman, R., Carey, V., Dudoit, S., Irizarry, R., Huber, W. (eds.) Bioinformatics and Computational Biology Solutions Using R and Bioconductor, pp. 397–420. Springer, New York (2005)
16. Davis, S., Meltzer, P.S.: GEOquery: a bridge between the Gene Expression Omnibus (GEO) and BioConductor. Bioinformatics 14, 1846–1847 (2007)
17. Benjamini, Y., Hochberg, Y.: Controlling the false discovery rate: a practical and powerful approach to multiple testing. Journal of the Royal Statistical Society, Series B 57(1), 289–300 (1995)
18. Ashburner, M., Ball, C.A., Blake, J.A., Botstein, D., Butler, H., Cherry, J.M., Davis, A.P., Dolinski, K., Dwight, S.S., Eppig, J.T., Harris, M.A., Hill, D.P., Issel-Tarver, L., Kasarskis, A., Lewis, S., Matese, J.C., Richardson, J.E., Ringwald, M., Rubin, G.M., Sherlock, G.: Gene ontology: tool for the unification of biology. The Gene Ontology Consortium 25(1), 25–29 (2000)
19. Xie, C., et al.: KOBAS 2.0: a web server for annotation and identi cation of enriched pathways and diseases. Nucleic Acids Research 39(suppl. 2), W316–W322 (2011)
20. Zhang, G., Li, J., Purkayastha, S., Tang, Y., Zhang, H., Yin, Y., Li, B., Liu, G., Cai, D.: Hypothalamic Programming of Systemic Aging Involving IKK, NF-kB and GnRH. Nature 497, 211–216 (2013)

Control of a Bioreactor
with Quantized Measurements

Francis Mairet and Jean-Luc Gouzé

INRIA BIOCORE, 2004 route des Lucioles,
BP 93, 06902 Sophia-Antipolis Cedex, France
{francis.mairet,jean-luc.gouze}@inria.fr

Abstract. We consider the problem of global stabilization of an unstable bioreactor model (e.g. for anaerobic digestion), when the measurements are discrete and in finite number ("quantized"), with control of the dilution rate. The model is a differential system with two variables, and the output is the biomass growth. The measurements define regions in the state space, and they can be perfect or uncertain (i.e. without or with overlaps). We show that a quantized control may lead to global stabilization: trajectories have to follow some transitions between the regions, until the final region where they converge toward the reference equilibrium. On the boundary between regions, the solutions are defined as a Filippov differential inclusion.

Keywords: bioreactor, Haldane model, hybrid systems, differential inclusions, quantized output, control.

1 Introduction

Classical control methods are often based on the complete knowledge of some outputs $y(t)$ of the system [21]. By complete, we mean that any output y_i is a real number, possibly measured with some noise δ_i. The control is then built with this (noisy) measurement. These tools have been successfully applied in many domains of science and engineering, e.g. in the domains of biosystems and bioreactors [10]. However, in these domains, detailed quantitative measurements are often difficult or impossible or too expensive. A striking example is the measurements of gene expression by DNA-chips, giving only a Boolean measure equal to on (gene expressed) or off (not expressed). In the domain of bioprocesses, it frequently happens that only a limited number or level of measurements are available (e.g. high, very high ...) because the devices only give a discretized semi-quantitative or qualitative measurement [6]. The measure may also be quantized by some physical device (as the time given by an analogical clock), and give as a result some number among a finite collection.

For this case of quantized outputs, the problem of control has also to be considered in a non-classical way: the control cannot be a function of the full continuous state variables anymore, and most likely will change only when the

F. Fages and C. Piazza (Eds.): FMMB 2014, LNBI 8738, pp. 47–62, 2014.

quantized measurement changes. Moreover, the control itself could be quantized, due to physical device limitations.

The above framework has been considered by numerous works, having their own specificity: quantized output and control with adjustable "zoom" and different time protocols, [19], hybrid systems abstracting continuous ones (cf. [17] for many examples of theories and applications).

In this paper, we consider a classical problem in the field of bioprocesses: the stabilization of an unstable bioreactor model, this model being a simplified representation of anaerobic fermentation, towards its working set point. Anaerobic fermentation is one of the most employed process for (liquid) waste treatment [10]. The process (substrate and biomass are state variables) has two stable equilibria (and an unstable one, with a separatrix between the two basins of attractions of the respective stable equilibria), one being the (undesirable) washout of the culture ([15]). The goal is to globally stabilize the process toward the other locally stable reference equilibrium. The (classical) output is the biomass growth (through gaseous production), the control is the dilution rate (see [22] for a review of control strategies). For scalar continuous output, there exists many approaches based on well-accepted models ([7]), using constant or adaptive yield [18,3]. Some original approaches make use of a supplementary competitor biomass [20].

In this paper, we suppose that the outputs are discrete or quantized: there are available in the form of finite discrete measurements. The precise models are described later: roughly, the simplest one is "perfect", without noise, meaning that the true measure is supposed to be one of the discrete measurements, and that the transitions between two contiguous discrete measures are perfectly known. The next model is an uncertain model where the discrete measurements may overlap, and the true value is at the intersection between two quantized outputs. Remark that the model of uncertainty is different from the interval observers approaches ([2,13]) for the estimation or regulation [1], where some outputs or kinetics are not well known, but upper or lower bounds are known. Moreover, in the interval observer case, the variables are classical continuous variables.

For this problem, the general approaches described above do not apply, and we have to turn to more tailored methods, often coming from the theory of hybrid systems, or quantized feedbacks (see above) ... We here develop our adapted "hybrid" approach. It has also some relations with the fuzzy modeling and control approach: see e.g. in a similar bioreactor process the paper [11]. We provide here a more analytic approach, and prove our results of stability with techniques coming from differential inclusions and hybrid systems theory [12]. Our work has some relations with theoretical qualitative control techniques used for piecewise linear systems in the field of genetic regulatory networks ([9]). The approach is also similar to the domain approaches used in hybrid systems theory, where there are some (controlled) transitions between regions, forming a transition graph [5,14].

The paper is organized as follows: the first section describes the bioreactor model and the measurements models, and gives useful elements for the following.

In the second section, we explain what happen at the boundary between two discrete measurements, and define the control on this boundary, with the help of the Filippov definition of differential inclusion. We follow by giving the full solution of the problem in some cases and examples, with or without uncertainty.

2 Framework

2.1 Model Presentation

In a perfectly mixed continuous reactor, the growth of biomass x limited by a substrate s can be described by the following system (see [4,10]):

$$\begin{cases} \dot{s} = u(t)(s_{in} - s) - k\mu(s)x \\ \dot{x} = (\mu(s) - u(t))x \end{cases} \tag{1}$$

where s_{in} is the input substrate concentration, $u(t)$ the dilution rate, k the pseudo yield coefficient, and $\mu(s)$ the specific growth rate.

Given $\xi = (s, x)$, let us rewrite System (1) as $\dot{\xi} = f(\xi, u(t))$, where the dilution rate $u(t)$ is the manipulated input.

The specific growth rate $\mu(s)$ is assumed to be a Haldane function (i.e. with substrate inhibition) [7]:

$$\mu(s) = \bar{\mu}\frac{s}{k_s + s + s^2/k_i} \tag{2}$$

Parameters $\bar{\mu}, k_s, k_i$ describe the model and are positive. This function admits a maximum for a substrate concentration $s = \sqrt{k_s k_i} := \bar{s}$, and we will assume $\bar{s} < s_{in}$.

Lemma 1. *The solutions of System* (1) *with initial conditions in the positive orthant are positive and bounded.*

Proof. It is easy to check that the solutions stay positive. Now consider $z = s + kx$ whose derivative writes

$$\dot{z} = u(t)(s_{in} - z).$$

It follows that z is upper bounded by $\max(z(0), s_{in})$, and so is kx. Finally, if $s(t) > s_{in}$, then $\dot{s}(t) < 0$, therefore s is upper bounded by $\max(s(0), s_{in})$.

In the following, we will assume initial conditions within the interior of the positive orthant.

2.2 Quantized Measurements

We consider that a growth proxy $y(\xi) = \alpha\mu(s)x$ of biomass growth is monitored (e.g. through gas production), but in a quantized way, in the form of a more or less qualitative measure: it can be levels (high, medium, low...) or discrete measures. Finally, we only know that $y(\xi)$ is in a given range, or equivalently that ξ is in a given region (parameter α is a positive yield coefficient):

$$Y_i = \{\xi \in \mathbb{R}^2_+ : \underline{y}_i \leq y(\xi) \leq \overline{y}_i\}, \ i = 1, \ldots n-1,$$
$$Y_n = \{\xi \in \mathbb{R}^2_+ : \underline{y}_n \leq y(\xi)\}.$$

where $0 = \underline{y}_1 < \underline{y}_2 < \ldots < \underline{y}_n$ and $\overline{y}_1 < \overline{y}_2 < \ldots < \overline{y}_{n-1}$. We will consider two cases:

– (A1). *Perfect* quantized measurements:

$$\overline{y}_i = \underline{y}_{i+1}, \ \forall i = \{1, \ldots n-1\}.$$

This corresponds to the case where there is no overlap between regions. The boundaries are perfectly defined and measured.

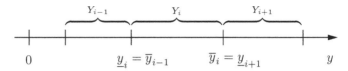

– (A2). *Uncertain* quantized measurements:

$$\underline{y}_i < \overline{y}_{i-1} < \underline{y}_{i+1}, \ \forall i = \{2, \ldots n-1\}.$$

In this case, we have overlaps between the regions. In these overlaps, the measure is not deterministic, and may belong to two values.

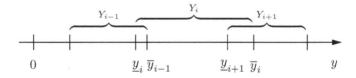

For both cases, we define (open) regular domains:

$$\tilde{Y}_i := Y_i \setminus (Y_{i-1} \cup Y_{i+1}),$$

and (closed) switching domains:

$$Y_{i|i+1} := Y_i \cap Y_{i+1}.$$

For perfect measurements (A1), we have $\tilde{Y}_i = intY_i$, and the switching domains $Y_{i|i+1}$ correspond to the lines $y(\xi) = \overline{y}_i = \underline{y}_{i+1}$. For uncertain measurements (A2), the switching domains $Y_{i|i+1}$ become the regions $\{\xi \in \mathbb{R}^2_+ : \underline{y}_{i+1} \leq y(\xi) \leq \overline{y}_i\}$.

In a switching domain $\xi \in Y_{i|i+1}$, we consider that the measurement is undetermined, i.e. either $\xi \in Y_i$ or $\xi \in Y_{i+1}$.

2.3   Quantized Control

Given the risk of washout, our objective is to design a feedback controller that globally stabilizes System (1) towards a set-point. Given that measurements are quantized, the controller should be defined with respect to each region:

$$\xi(t) \in Y_i \iff u(t) = D_i, \quad i = 1, ..., n. \tag{3}$$

Here D_i is the positive dilution rate in region i. This control scheme leads to discontinuities in the vector fields. Moreover, in the switching domains, the control is undetermined. Thus, solutions of System (1) under Control law (3) are defined in the sense of Filippov, as the solutions of the differential inclusion [12]:

$$\dot{\xi} \in H(\xi)$$

where $H(\xi)$ is defined on regular domains $\tilde{Y}_i$ as the ordinary function $H(\xi) = f(\xi, D_i)$, and on switching domains $Y_{i|i+1}$ as the closed convex hull of the two vector fields in the two domains i and $i+1$:

$$H(\xi) = \overline{co}\{f(\xi, D_i), f(\xi, D_{i+1})\}.$$

Following [8,9], a solution of System (1) under Control law (3) on $[0, T]$ is an absolutely continuous (w.r.t. t) function $\xi(t, \xi_0)$ such that $\xi(t, \xi_0) = \xi_0$ and $\dot{\xi} \in H(\xi)$ for almost all $t \in [0, T]$.

3   Model Analysis with a Constant Dilution

For this case, the system is a classical ordinary differential equation in the whole space. When a constant dilution rate D is applied (i.e. $u(t) = D$, $\forall t \geq 0$), System (1) can present bistability, with a risk of washout. Let us denote $s_a(D)$ and $s_b(D)$ the two solutions, for $D \in (0, \mu(\bar{s}))$, of the equation $\mu(s) = D$, with $0 < s_a(D) < \bar{s} < s_b(D)$. For the Haldane growth rate defined by (2), we have:

$$s_a(D) = \frac{k_i}{2}\left(\frac{\bar{\mu}}{D} - 1\right) - \sqrt{\left(\frac{k_i}{2}\left(\frac{\bar{\mu}}{D} - 1\right)\right)^2 - k_s k_i}$$

$$s_b(D) = \frac{k_i}{2}\left(\frac{\bar{\mu}}{D} - 1\right) + \sqrt{\left(\frac{k_i}{2}\left(\frac{\bar{\mu}}{D} - 1\right)\right)^2 - k_s k_i}.$$

The asymptotic behavior of the system can be summarized as follows:

Proposition 1. *Consider System (1) with a constant dilution rate $u(t) = D$ and initial conditions in the interior of the positive orthant.*
(i) If $D < \mu(s_{in})$, the system admits a globally exponentially stable equilibrium $\xi_a(D) = \left(s_a(D), \frac{s_{in} - s_a(D)}{k}\right)$.
(ii) If $\mu(s_{in}) < D < \mu(\bar{s})$, the system admits two locally exponentially stable equilibria, a working point $\xi_a(D) = \left(s_a(D), \frac{s_{in} - s_a(D)}{k}\right)$ and the washout $\xi_0 = (s_{in}, 0)$, and a saddle point $\xi_b(D) = \left(s_b(D), \frac{s_{in} - s_b(D)}{k}\right)$, see Figure 1.
(iii) If $D > \mu(\bar{s})$, the washout $\xi_0 = (s_{in}, 0)$ is globally exponentially stable.

Proof. See [15].

For $D \in (0, \mu(\bar{s}))$, let us define:

$$y_j(D) = \frac{\alpha D}{k}[s_{in} - s_j(D)], \quad j = a, b.$$

$y_a(D)$ and $y_b(D)$ are the growth proxy obtained respectively at the equilibria $\xi_a(D)$ and $\xi_b(D)$ (if it exists)[1].

In order to design our control law, we need to provide some further properties of the system dynamics. In particular, we need to characterize $\dot{y}(\xi)$, the time derivative of $y(\xi)$ along a trajectory of System (1) with a constant dilution rate $u(t) = D$:

$$\dot{y}(\xi) = \alpha\left[D(s_{in} - s) - k\mu(s)x\right]\mu'(s)x + \alpha\mu(s)(\mu(s) - D)x.$$

Let us define the following functions:

$$g_D : s \longmapsto \frac{\mu(s) - D}{k\mu'(s)} + \frac{D(s_{in} - s)}{k\mu(s)}$$

$$h_D^j : s \longmapsto \frac{D(s_{in} - s_j(D))}{k\mu(s)}, \quad j = a, b.$$

In the (s, x) plane, $g_D(s)$, $h_D^a(s)$ and $h_D^b(s)$ represent respectively the nullcline $\dot{y}(\xi) = 0$ and the isolines $y(\xi) = y_a(D)$ and $y(\xi) = y_b(D)$ (i.e. passing through the equilibria $\xi_a(D)$ and $\xi_b(D)$), see Figure 1. Knowing that the nullcline $\dot{y}(\xi) = 0$ is tangent to the isoline $y(\xi) = y_a(D)$ (resp. $y(\xi) = y_b(D)$) at the equilibrium point $\xi_a(D)$ (resp. $\xi_b(D)$), we will determine in the next lemma the relative positions of these curves, see Fig. 1.

Lemma 2. *(i) For $s \in (0, \bar{s})$, we have $g_D(s) \geq h_D^a(s)$: the nullcline $\dot{y}(\xi) = 0$ is over the isoline $y(\xi) = y_a(D)$*
(ii) For $s \in (\bar{s}, s_{in})$, we have $g_D(s) \leq h_D^b(s)$: the nullcline $\dot{y}(\xi) = 0$ is below the isoline $y(\xi) = y_b(D)$.

Proof. See Appendix.

This allows us to determine the monotonicity of $y(\xi)$ in a region of interest (for the design of the control law).

Lemma 3. *Consider System (1) with a constant dilution rate $u(t) = D$. For $\xi \in \mathbb{R}_+^2$ such that $y_b(D) < y(\xi) < y_a(D)$, we have $\dot{y}(\xi) > 0$.*

Proof. See Appendix.

---

[1] If $D < \mu(s_{in})$, $\xi_b(D)$ does not exist and $y_b(D) < 0$.

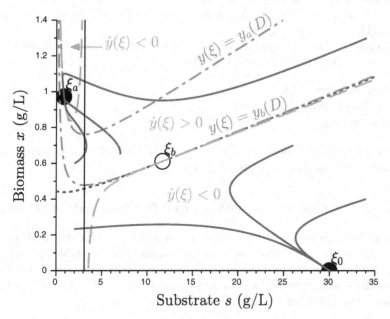

Fig. 1. Phase portrait of System (1) with a constant dilution rate $u(t) = D \in (\mu(s_{in}); \mu(\bar{s}))$. Magenta lines: trajectories, cyan dashed lines: nullcline $\dot{y}(\xi) = 0$, green dash dotted lines: isolines $y(\xi) = y_a(D)$ and $y(\xi) = y_b(D)$, red dotted line: separatrix, black vertical line: $s = \bar{s}$, dark circles: stable equilibria, open circle: unstable equilibrium.

4   Control with Quantized Measurements

4.1   Control Design

Our goal is to globally stabilize the system towards the stable equilibrium corresponding to D_n. Number n is the number of measurements (see section 2.2) and the control is such that $D_1 < D_2 < \ldots < D_n$. The last dilution rate D_n is chosen because of its high productivity, it verifies the case (ii) of Prop. 1. We consider the following control law, based on the quantized measurements $y(\xi)$:

$$\forall t \geq 0, \quad \xi(t) \in Y_i \Leftrightarrow u(\xi) = D_i, \tag{4}$$

given that the following conditions are fulfilled:

$$y_b(D_i) < \underline{y}_i \quad i = 1, \ldots, n, \tag{5}$$

$$y_a(D_i) > \overline{y}_i \quad i = 1, \ldots, n-1, \tag{6}$$

$$y_a(D_n) > \overline{y}_{n-1}. \tag{7}$$

These conditions make the equilibrium $\xi_a(D_n)$ globally stable, as we will see below. In Section 5.1, we will precise how to choose the D_i such that these conditions hold.

In order to prove the asymptotic behavior of System (1) under Control law (4-7), the study will be divided into three steps:

- the dynamics in one region,
- the transition between two regions,
- the global dynamics.

This approach is similar to those deducing the global dynamics from a "transition graph" of possible transitions between regions.

4.2   Dynamics in One Region with a Given Dilution: Exit of Domain

We first focus on a region Y_i, $i < n$. A constant dilution D_i - such that Conditions (5-6) for i hold - is applied. These conditions guarantee that the stable operating equilibrium for this dilution (see Proposition 1) is located in an upper region Y_j, $j > i$, while the saddle point is located in a lower region Y_k, $k < i$. This allows us to establish the following lemma:

Lemma 4. *For any $i \in \{1, ..., n-1\}$, consider System (1) under a constant control $u(t) = D_i$, such that Conditions (5-6) for i hold. All solutions with initial conditions in Y_i leaves this domain, crossing the boundary $y(\xi) = \overline{y}_i$.*

Proof. Let us consider the function $V(\xi) = y_a(D_n) - y(\xi)$ on Y_i. Given Conditions (5-6), we get $\forall \xi \in Y_i$:

$$y_b(D_i) < \underline{y}_i < y(\xi) < \overline{y}_i < y_a(D_i).$$

Since a constant dilution rate D_i is applied, we can apply Lemma 3 to conclude that $\dot{y}(\xi) > 0$.

Thus, $V(\xi)$ is decreasing on Y_i. Recalling that the trajectories are also bounded (Lemma 1), we can apply LaSalle invariance theorem [16] on the domain $\Omega_1 := \{\xi \in Y_i \mid x \le \max(z(0), s_{in})/k, \ s \le \max(s(0), s_{in})\}$. Given that the set of all the points in Ω_1 where $\dot{V}(\xi) = 0$ is empty, any trajectory starting in Ω_1 will leave this region. The boundaries $x = \max(z(0), s_{in})/k$ and $s = \max(s(0), s_{in})$ are repulsive (see Proof of Lemma 1). Finally, the boundary $y(\xi) = \underline{y}_i$ corresponds to the maximum of $V(\xi)$ on Ω_1, so every trajectory will leaves this domain, crossing the boundary $y(\xi) = \overline{y}_i$.

4.3   Transition between Two Regions

Now we will characterize the transition between regions (as we have seen above, the intersection can be either a simple curve in the case of perfect measurements, or a region with non empty interior in the uncertain case):

Lemma 5. *For any $i \in \{1, ..., n-1\}$, consider System (1) under Control law (4) with Conditions (5-6) for $i, i+1$[2]. All trajectories with initial conditions in $\tilde{Y}_i \cup Y_{i|i+1}$ enter the regular domain $\tilde{Y}_{i+1}$.*

Proof. First, we consider $i \neq n-1$. We will follow the same reasoning as for the previous lemma, applying Lasalle theorem on a domain

$$\Omega_2 := \{\xi \in int\mathbb{R}_+^2 \mid x \leq \max(z(0), s_{in})/k, \ s \leq \max(s(0), s_{in}), \ \overline{y}_{i-1} < y(\xi) < y^\dagger\}$$

with $\overline{y}_i < y^\dagger < \underline{y}_{i+2}$. We can show that the functional $V(\xi) = y_a(D_n) - y(\xi)$ is decreasing on Y_i whenever $u = D_i$. Similarly, $V(\xi)$ is also decreasing on Y_{i+1} whenever $u = D_{i+1}$. Now under Control law (4), we have shown that $V(\xi)$ is decreasing on the regular domains $\tilde{Y}_i$ and $\tilde{Y}_{i+1}$. $V(\xi)$ is a regular C^1 function, and can be differentiated along the differential inclusion. On the switching domains $Y_{i|i+1}$, we have:

$$\dot{V}(\xi) \in \overline{co}\left\{\frac{\partial V}{\partial x}f(\xi, D_i), \frac{\partial V}{\partial x}f(\xi, D_{i+1})\right\} < 0.$$

Thus, $V(\xi)$ is decreasing on Ω_2. Following the proof of Lemma 4 concerning the boundaries, we can deduce that every trajectory will reach the boundary $y(\xi) = y^\dagger$, i.e. it will enter $\tilde{Y}_{i+1}$.
For $i = n-1$, taking $\overline{y}_{n-1} < y^\dagger < y_a(D_n)$, we can show similarly that $V(\xi)$ is decreasing on Ω_2 so every trajectory will enter the region $\tilde{Y}_n$.

Following the same proof, we can show that the reverse path is not possible, in particular for the last region:

Lemma 6. *Consider System (1) under Control law (4) with Conditions (5) for $n-1, n$, Condition (6) for $n-1$, and Condition (7). The regular domain $\tilde{Y}_n$ is positively invariant.*

4.4   Global Dynamics

Proposition 2. *Control law (4-7) globally stabilizes System (1) towards the point $\xi_a(D_n)$.*

Proof. From Lemmas 5 and 6, we can deduce that every trajectory will enter the regular domain $\tilde{Y}_n$, and that this domain is positively invariant.
   System (1) under a constant control $u(t) = D_n$ has two non-trivial equilibria (see Proposition 1): $\xi_a(D_n)$, and $\xi_b(D_n)$. The growth proxy at these two points satisfy $y_a(D_n) > \overline{y}_{n-1}$ and $y_b(D_n) < \underline{y}_n$ (Conditions (5,7)), so there is only one equilibrium in $\tilde{Y}_n$: $\xi_a(D_n)$. Moreover, it is easy to check that $\tilde{Y}_n$ is in the basin of attraction of $\xi_a(D_n)$, therefore all trajectories will converge toward this equilibrium.

---

[2] Or if $i = n-1$, Conditions (5) for $n-1, n$, Condition (6) for $n-1$, and Condition (7).

5   Implementation of the Control Law

5.1   How to Fulfill Conditions (5-7)

The global stability of the control law is based on Conditions (5-7). We consider the case where the regions are imposed (by technical constraints of the measurements) and we want to find the different dilution rates D_i such that these conditions hold. The approach may be analytic or graphical, and will described elsewhere.

For example, for perfect measurements (A1) with equidistribution, it can be shown that it will always be possible to implement the desired control law, i.e. it is always possible to find a set of dilution rates D_i such that Conditions (5-7) hold, whenever the measurement resolution is good enough (i.e. the number of regions is high enough). The result is illustrated by the simulations below.

5.2   Simulations

As an example, we consider the anaerobic digestion process, where the methane production rate is measured. Parameters, given in Table 1, are inspired from [7] (considering only the methanogenesis step). Our objective is to stabilize the equilibrium $\xi_a(D^*)$, with $D^* = 0.47$ d^{-1} (which corresponds to a productivity of 92% of the maximal productivity).

Table 1. Parameter values used for simulation

Parameter	Value
$\bar{\mu}$	0.74 d^{-1}
k_s	0.59 g.L^{-1}
k_i	16.4 g.L^{-1}
k	30
α	11 L CH$_4$.g^{-1}
s_{in}	30 g.L^{-1}

For uncertain measurements, we use discrete time simulation. At each time step t_k (with $\Delta t = 0.05$d), when $\xi(t_k)$ is in a switching region $Y_{i|i+1}$, we choose randomly the control $u(t_k)$ between D_i and D_{i+1}. In this case, we perform various simulations for a same initial condition.

Trajectories for various initial conditions are represented in the phase portrait for perfect and uncertain measurements, see Figure 2 and Figure 3.

On top of both figures, the number of regions (three regions only) is too small in order to stabilize the given set-point $\xi_a(D^*)$: it is not possible to choose dilution rates such that Conditions (5-7) hold. In this case, some trajectories do not converge towards the set-point. Some regions have transitions towards the upper region, but also towards the lower one. There are sliding modes. This aspect will be further discussed in the next subsection.

On bottom of both figures, with one more region (four measurements), we can define a set of dilution rates such that Conditions (5-7) are fulfilled. The transition graph is deterministic (there is only one transition from a region to the upper one).

Actually, given the following perfect measurement set (considering equidistant region):

$$\overline{y}_i = \underline{y}_{i+1} = \frac{i}{n-1}\overline{y}_n, \quad i = 1, ..., n-1, \quad \text{with} \quad \overline{y}_n = 4 \text{ L CH}_4.\text{L}^{-1}.\text{d}^{-1},$$

we can choose the following dilution rates:

$$D_1 = 0.19 \text{ d}^{-1}, \quad D_2 = 0.29 \text{ d}^{-1}, \quad D_3 = 0.4 \text{ d}^{-1}, \quad D_4 = 0.47 \text{ d}^{-1}.$$

For uncertain measurements, we increased each upper bound and decreased each lower bound by 10%. It appears that the same dilution rates can be chosen.

Thus, all the trajectories converge towards the set-point (see Figure 3).

5.3   When Conditions Are Not Verified: Risk of Failure

We here detail what happens if Conditions (5-7) are not fulfilled, and in particular if there is a risk of washout. This point is illustrated by the top figures of Fig. 2 and Fig. 3.

First, given the previous analysis of the system, one can easily see that only the condition $y_b(D_1) < \underline{y}_1$, i.e. $D_1 < \mu(s_{in})$ is necessary to prevent a washout, so D_1 can be chosen with a safety margin in order to avoid such situation. Now, if Condition (6) does not hold for some i, the stable equilibrium $\xi_a(D_i)$ will be located in the region Y_i, so some trajectories can converge towards this point instead of going to the next region. On the other hand, if Condition (5) is not fulfilled for some $i > 1$, the unstable equilibrium $\xi_b(D_i)$ will be located in the region Y_i, and thus a trajectory can stay in a switching domain. Given that z converges towards s_{in}, such trajectory will converge towards the intersection between the switching domain and the invariant manifold $z = s_{in}$ (see Figure 2, Figure 3 on top):

- For perfect measurements (A1), this gives rise to a sliding mode and the convergence towards a singular equilibrium point.
- For uncertain measurements (A2), all the trajectories converge towards a line segment. In our simulation, they actually also converge towards a singular equilibrium point.

In all the cases, the trajectories converge towards a point or a line segment. Although it is not desired, this behavior is particularly safe (given that there is theoretically no risk of washout). Moreover, undesired equilibrium can easily be detected and the dilution rates can be changed accordingly (manually or through a supervision algorithm).

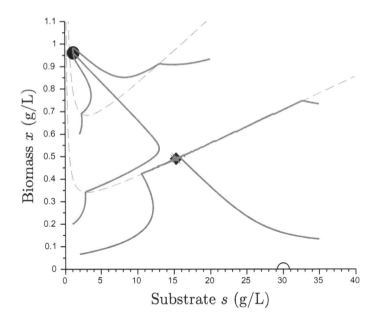

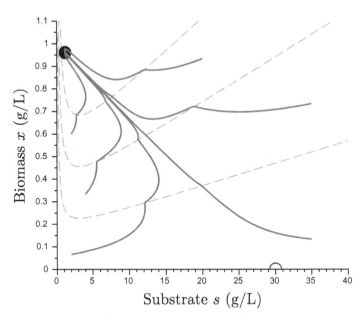

Fig. 2. Trajectories (magenta lines) with Control law (4) for various initial conditions in the phase portrait, in the case of perfect measurements. Top: Conditions (5-7) are not fulfilled, some trajectories converge towards a singular equilibrium point (black diamond) with a sliding mode. Bottom: Conditions (5-7) are fulfilled, all the trajectories converge towards the set-point (dark circle). Open circle: washout. The frontiers are represented by the green dashed lines.

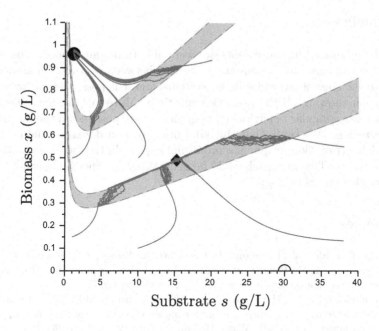

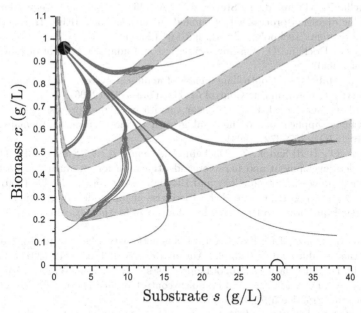

Fig. 3. Trajectories (magenta lines) with Control law (4) for various initial conditions in the phase portrait, in the case of uncertain measurements. Top: Conditions (5-7) are not fulfilled, some trajectories converge towards a singular equilibrium point (black diamond) with a sliding mode. Bottom: Conditions (5-7) are fulfilled, all the trajectories converge towards the set-point (dark circle). The frontiers are represented by the green dashed lines, switching regions are colored in gray. In these gray regions, the system is not deterministic. Open circle: washout.

6 Conclusion

Given the quantized measurements, we were able to design (under some conditions) a control based on regions and transition between regions. These tools are similar to the ones of piecewise linear systems, and it is possible to draw a transition graph showing all the possible transitions. Moreover, we have seen that for some cases, singular behaviors (sliding modes) are possible on the boundaries between regions. We think that this kind of control on domains, and design of the resulting transition graph, is a promising approach, that we want to deepen in future works. This approach could be generalized to others classical systems, e.g. in mathematical ecology.

References

1. Alcaraz-Gonzalez, V., Harmand, J., Rapaport, A., Steyer, J.P., Gonzalez-Alvarez, V., Pelayo-Ortiz, C.: Robust interval-based regulation for anaerobic digestion processes. Water Science & Technology 52(1-2), 449–456 (2005)
2. Alcaraz-González, V., Harmand, J., Rapaport, A., Steyer, J.P., Gonzalez-Alvarez, V., Pelayo-Ortiz, C.: Software sensors for highly uncertain wwtps: a new approach based on interval observers. Water Research 36(10), 2515–2524 (2002)
3. Antonelli, R., Harmand, J., Steyer, J.P., Astolfi, A.: Set-point regulation of an anaerobic digestion process with bounded output feedback. IEEE Transactions on Control Systems Technology 11(4), 495–504 (2003)
4. Bastin, G., Dochain, D.: On-line estimation and adaptive control of bioreactors. Elsevier (1990)
5. Belta, C., Habets, L.: Controlling a class of nonlinear systems on rectangles. IEEE Transactions on Automatic Control 51(11), 1749–1759 (2006)
6. Bernard, O., Gouzé, J.-L.: Non-linear qualitative signal processing for biological systems: application to the algal growth in bioreactors. Mathematical Biosciences 157(1), 357–372 (1999)
7. Bernard, O., Hadj-Sadok, Z., Dochain, D., Genovesi, A., Steyer, J.P.: Dynamical model development and parameter identification for an anaerobic wastewater treatment process. Biotechnology and Bioengineering 75(4), 424–438 (2001)
8. Casey, R., de Jong, H., Gouzé, J.-L.: Piecewise-linear models of genetic regulatory networks: Equilibria and their stability. Journal of Mathematical Biology 52, 27–56 (2006)
9. Chaves, M., Gouzé, J.L.: Exact control of genetic networks in a qualitative framework: the bistable switch example. Automatica 47(6), 1105–1112 (2011)
10. Dochain, D.: Automatic control of bioprocesses, vol. 28. John Wiley & Sons (2010)
11. Estaben, M., Polit, M., Steyer, J.P.: Fuzzy control for an anaerobic digester. Control Engineering Practice 5(9), 1303–1310 (1997)
12. Filippov, A.F.: Differential Equations with Discontinuous Righthand Sides. Kluwer Academic Publishers, Dordrecht (1988)
13. Gouzé, J.L., Rapaport, A., Hadj-Sadok, M.Z.: Interval observers for uncertain biological systems. Ecological Modelling 133(1), 45–56 (2000)
14. Habets, L., van Schuppen, J.H.: A control problem for affine dynamical systems on a full-dimensional polytope. Automatica 40(1), 21–35 (2004)

15. Hess, J., Bernard, O.: Design and study of a risk management criterion for an unstable anaerobic wastewater treatment process. Journal of Process Control 18(1), 71–79 (2008)
16. LaSalle, J.P.: The Stability of Dynamical Systems. CBMS-NSF Regional Conference Series in Applied Mathematics. Society for Industrial and Applied Mathematics (1976)
17. Lunze, J., Lamnabhi-Lagarrigue, F.: Handbook of hybrid systems control: theory, tools, applications. Cambridge University Press (2009)
18. Mailleret, L., Bernard, O., Steyer, J.P.: Nonlinear adaptive control for bioreactors with unknown kinetics. Automatica 40(8), 1379–1385 (2004)
19. Nesic, D., Liberzon, D.: A unified framework for design and analysis of networked and quantized control systems. IEEE Transactions on Automatic Control 54(4), 732–747 (2009)
20. Rapaport, A., Harmand, J.: Biological control of the chemostat with nonmonotonic response and different removal rates. Mathematical Biosciences and Engineering 5(3), 539–547 (2008)
21. Sontag, E.D.: Mathematical control theory: deterministic finite dimensional systems, vol. 6. Springer (1998)
22. Steyer, J.P., Bernard, O., Batstone, D.J., Angelidaki, I.: Lessons learnt from 15 years of ica in anaerobic digesters. Instrumentation, Control and Automation for Water and Wastewater Treatment and Transport Systems IX 53(4), 25–33 (2006)

7   Appendix

Proof of Lemma 2

Let us define $\varphi_D^j(s) := g_D(s) - h_D^j(s)$ for $j = a, b$. We have:

$$\varphi_D^j(s) = \frac{D(s_j(D) - s)}{k\mu(s)} + \frac{\mu(s) - D}{k\mu'(s)},$$

$$\varphi_D^j{}'(s) = \frac{1}{k}(\mu(s) - D)\left(\frac{1}{\mu(s)} - \frac{\mu''(s)}{\mu'(s)^2}\right) - D(s_j(D) - s)\frac{\mu'(s)}{k\mu(s)^2}.$$

First, we consider $\varphi_D^a(s)$ on $(0, \bar{s})$. Given that $\mu(s)$ is increasing and concave on this interval, we get $\varphi_D^a{}'(s) < 0$ on $(0, s_a(D))$, and $\varphi_D^a{}'(s) > 0$ on $(s_a(D), \bar{s})$. Moreover, we have $\varphi_D^a(s_a(D)) = 0$, so $\varphi_D^a(s) \geq 0$ on $(0, \bar{s})$, which proves (i).

Now we want to determine the sign of $\varphi_D^b(s)$ on $(\bar{s}, +\infty)$. For this purpose, we consider the equation $\varphi_D^b(s) = 0$. By replacing $\mu(s)$ and its derivative by their analytic expressions, this equation becomes:

$$\frac{s_a(D)}{k_i}s^2 - 2k_s s + k_s s_b(D) = 0.$$

Given that $s_a(D)s_b(D) = k_s k_i$, the equation $\varphi_D^b(s) = 0$ has only one root $s = s_b(D)$. Moreover, we have:

$$\lim_{s \searrow \bar{s}} \varphi_D^b(s) = -\infty \quad \text{and} \quad \lim_{s \to +\infty} \varphi_D^b(s) = -\infty.$$

Given that $\varphi_D^b(s)$ is continuous on $(\bar{s}, +\infty)$, we finally conclude that on this interval, $\varphi_D^b(s) \leq 0$, i.e. $g_D(s) \leq h_D^b(s)$. $\qquad\square$

Proof of Lemma 3

First, given that $g_D(s)$ represent the nullcline $\dot{y}(\xi) = 0$, we can check that we have (see Figure 1):

$$\dot{y}(\xi) > 0 \text{ on } \{\xi \in int\mathbb{R}_+^2 \mid s < \bar{s}, x < g_D(s)\} \cup \{\xi \in int\mathbb{R}_+^2 \mid s > \bar{s}, x > g_D(s)\}.$$

Recalling that $h_D^a(s)$ and $h_D^b(s)$ are respectively the isolines $y(\xi) = y_a(D)$ and $y(\xi) = y_b(D)$, Lemma 2 allows to conclude that for $\xi \in \mathbb{R}_+^2$ such that $y_b(D) < y(\xi) < y_a(D)$, we have $\dot{y}(\xi) > 0$. $\square$

External Interactions
on Hybrid Models of Biological Systems

Alberto Casagrande[1] and Carla Piazza[2]

[1] Dept. of Mathematics and Geosciences, University of Trieste, Italy
acasagrande@units.it
[2] Dept. of Mathematics and Computer Science, University of Udine, Italy
carla.piazza@uniud.it

Abstract. We propose a general framework for the analysis of hybrid automata representing biological systems which interacts with an environment. Our framework is based on unwinding conditions and it aims at establishing which external interactions substantially change the system behaviours. We exploit our proposal for the analysis of influenza disease treatable with both antivirals and interferons.

Keywords: Hybrid Automata Unwinding Behavioural Equivalences.

Introduction

Hybrid systems mix a discrete control and continuous dynamics. They evolve according to a continuous law which is updated in a discrete way by a finite collection of triggers. Hybrid automata were introduced to model hybrid systems (e.g., see [32,1]). They are finite state automata, equipped with continuous variables, whose states are labeled by dynamical systems (e.g., differential equations). The values of the variables are ruled by the dynamical system labeling the active state and, whenever an activation region is reached, the automaton crosses a discrete edge and changes its dynamics. This formalism has been widely used to represent biological models (e.g., see [18,8,10]).

Temporal Logics [36,37] and Model Checking [11] are standard tools for the analysis of hybrid automata (e.g., see [1,24]). The former provide specification languages for formulating the properties of interest. The latter furnishes algorithms for verifying properties on models. Two models are considered behaviorally equivalent when they are indistinguishable with respect to the temporal logic formulæ. Behavioral equivalences can then be used to both compare models and reduce their sizes. While their use as reduction criteria is well known in Model Checking, where the state explosion problem is a major concern, their importance for comparing models has been fruitfully exploited in other fields. In particular, in Information Flow Security behavioral equivalences are at the basis of unwinding conditions which allow to establish whether the system would behave correctly in hostile environments (see [40,5]).

In this paper, we model both biological systems and environments by using hybrid automata and we represent their interaction as a composition of their models. In this setting, we propose a robustness definition based on unwinding conditions that tries to

F. Fages and C. Piazza (Eds.): FMMB 2014, LNBI 8738, pp. 63–81, 2014.

match the behavior of a system and that of the same system into a (possibly hostile) environment: whenever they exhibit the same behaviors the system is robust with respect to that environment.

We show that the proposed techniques can be applied to investigate an original model developed to represent the influenza kinetics. The model consists of an hybrid automaton whose discrete states represent different treatments. In particular, it describes both antiviral and interferon based treatments. We establish the effectiveness of the two treatments on the virus load (number of viruses) and we observe the differences between them. The automaton has been obtained by integrating two models described in the literature ([23,42]) and by fitting its parameters on simulated data.

The use of the notions of *environment* and *composition* makes our framework particularly appealing for those applications of macro-biology where interactions play a central role.

The paper is organized as follows: Section 1 introduces multilevel labeled transition systems and hybrid automata and proposes them as a formalism to represents both biological systems and attacking environments. In Section 2, we use the notion of composition of hybrid automata to model the interactions between a biological system and an environment. Section 3 presents unwinding conditions as a formalization of robustness and briefly compares it with the related literature. In Section 4, we present three different unwinding conditions over hybrid automata and we show characterize one of these in terms of composition with any possible environment. Such unwinding conditions are then are exploited on kinetic models of influenza virus in Section 5. Conclusions are drawn in Section 6.

1 Preliminaries

1.1 Multilevel Labeled Transition Systems

The operational semantics of our models is given in terms of *labeled transition systems* (LTS). An LTS is a directed graph with labels on edges. Since we will model biological systems which should preserve part of their behaviors in any (possibly hostile) environment we will use *multilevel labeled transition systems (MLTS)*. In such labeled transition systems edge labels are partitioned into *Imposed* labels (*Imp*) and *Exposed* labels (*Exp*). Imposed labels represent the interactions with the environment on which the system has low or no control. An interaction occurs when an imposed action ι modeling an input synchronizes with its correspondent output $\bar{\iota}$. Such synchronization generates a silent transition labeled with the special label τ. Exposed labels represent the behaviors which should not be influenced by the environment, i.e., what the system exposes.

Given a set of labels A we use the notation $\overline{A}$ to denote the set of labels $\overline{A} = \{\bar{a}|a \in A\}$. From now on we implicitly assume that τ is neither an imposed nor an exposed label.

Definition 1 (Multilevel Labelled Transition System). *A Multilevel Labeled Transition System (MLTS) is a tuple* $(V, V_I Imp \cup Exp, E)$ *where:*

- *V is a set of nodes and $V_I \subseteq V$ is a not-empty set of initial nodes;*
- *Imp $\cup$ Exp is the set of input edge labels, where Imp are the imposed labels and Exp are the exposed ones;*

- $E \subseteq V \times (Imp \cup \overline{Imp} \cup \mathcal{E}xp \cup \overline{\mathcal{E}xp} \cup \{\tau\}) \times V$ *is a set of edges.*

We write $v \xrightarrow{\alpha} v'$ to indicate that (v, α, v') is an edge, i.e., $(v, \alpha, v') \in E$, and $v \to v'$ to denote that there exists $\alpha \in Imp \cup \overline{Imp} \cup \mathcal{E}xp \cup \overline{\mathcal{E}xp} \cup \{\tau\}$ such that $(v, \alpha, v') \in E$. We use ι to denote a generic element of $Imp \cup \overline{Imp}$ and ϵ for a generic element of $\mathcal{E}xp \cup \overline{\mathcal{E}xp}$. In some cases, also labels on nodes can be introduced.

MLTS's have been widely used in the area of *information flow security* to represent multilevel systems in which confidential (high-level) and public (low-level) data coexist. In such a context one should ensure that no information flow from a level to a lower one or, in other terms, that the high-level behaviors do not influence the low-level ones (see, e.g., [21,41,16].) Our biological framework takes inspiration from information flow security: *Imp* transitions play the role of "high-level" actions, while *$\mathcal{E}xp$* transitions represent the "low-level" ones.

Once a MLTS representing a system has been obtained, behavioral equivalences can be used to reduce the size of the MLTS, prove properties over the system, and compare different systems. Behavioral equivalences are equivalences over MLTS's that relate nodes having "similar behaviors". *Trace equivalence* and *bisimulation* are two of the most used behavioral equivalences in the literature. Trace equivalence relates two nodes if and only if they generate the same sequences of edge labels (traces). It produces a drastic reduction of the model size, but its computation is expensive (PSPACE-complete). On the contrary, bisimulation is a finer relation which is easier to compute. In its strong version it equates models that satisfy exactly the same formulæ of modal and branching temporal logics (see, e.g., [11]).

Definition 2 (Strong Bisimulation). *Given a MLTS $T = (V, V_I, Imp \cup \mathcal{E}xp, E)$, a strong bisimulation over T is a symmetric relation $R \subseteq V \times V$ such that for each $(u, v) \in R$ the following conditions hold:*

- $u \in V_I$ *if and only if $v \in V_I$;*
- *if $u \xrightarrow{\alpha} u'$, then $v \xrightarrow{\alpha} v'$ and $(u', v') \in R$.*

Two nodes $u, v \in V$ are said to be strongly bisimilar, *denoted as $u \sim v$, if there exists a strong bisimulation R such that $(u, v) \in R$.*

The relation $\sim$ is an equivalence relation, it is the largest strong bisimulation relation, and Paige-Tarjan algorithm [35] computes it in time $O(|E| \log |V|)$. Moreover, $u \sim v$ implies that u and v are trace-equivalent.

There exist many variant of bisimulation, depending on the applications one is interested in. Weak bisimulation is one of such variants and it is useful when synchronizations are modeled as internal silent transitions that cannot be observed. In these cases $v \xRightarrow{\alpha} v'$ denotes that $v(\xrightarrow{\tau})^* \xrightarrow{\alpha} (\xrightarrow{\tau})^* v'$, where $(\xrightarrow{\tau})^*$ is a possibly empty sequence of τ transitions. Moreover, $v \xRightarrow{\hat{\alpha}} v'$ denotes that $v \xRightarrow{\alpha} v'$, if α is different from τ, while $v(\xrightarrow{\alpha})^* v'$, if $\alpha = \tau$.

Definition 3 (Weak Bisimulation). *Given a MLTS $T = (V, V_I, Imp \cup \mathcal{E}xp, E)$, a weak bisimulation over T is a symmetric relation $R \subseteq V \times V$ such that for each $(u, v) \in R$ the following conditions hold:*

- $u \in V_I$ *if and only if* $v \in V_I$;
- *if* $u \xrightarrow{\alpha} u'$, *then* $v \xRightarrow{\hat{a}} v'$ *and* $(u', v') \in R$.

Two nodes $u, v \in V$ *are said to be* weakly bisimilar, *denoted as* $u \approx v$, *if there exists a weak bisimulation R such that* $(u, v) \in R$.

Definition 3 provides the standard notion of weak bisimulation, as given for instance in CCS [33], in which a silent transition can be simulated through an empty sequence of silent transitions. The relation $\approx$ is the largest weak bisimulation and it is an equivalence relation. It can be computed in time $O(|V| \|E|)$ by first computing the MLTS whose edges are $\Rightarrow$, then applying Paige-Tarjan algorithm.

Both the above notions of bisimulation can be easily defined also as relations between nodes of different MLTS's provided that they share the same sets $Imp \cup Exp$ of edge labels. In the rest of this paper we will use bisimulations in both sense. In particular, we will use the notation $T_1 \approx T_2$ to denote that there exists a weak bisimulation R such that for each node v_1 of T_1 there exists at least one node v_2 of T_2 such that $(v_1, v_2) \in R$ and vice-versa.

1.2 Hybrid Automata

Hybrid models, such as hybrid automata, mix discrete and continuous dynamics. Many different definitions of hybrid automaton have been suggested in the literature [1,24,32,7]. In this paper, we rely on a definition which is close to the one given by Lynch *et al.* [31] as the unwinding framework we will propose fits into it in a natural way.

Definition 4 (Hybrid Automata). *A* hybrid automaton *(HA)* $\mathcal{A} = (Imp, Exp, Q, \Theta, Imp, Exp^-, D, \mathcal{T})$ *consists of:*

- *Two disjoint sets of variables: a set Imp of* imposed *variables and a set Exp of* exposed *variables.* $Var \overset{def}{=} Imp \cup Exp$;
- *A set Q of values, called* states, *that Var can assume;*
- *A nonempty set* $\Theta \subseteq Q$ *of start states;*
- *Two disjoint sets of actions: Imp (*imposed *discrete actions) and Exp^- (*exposed *discrete actions). Both of them does not contain the action τ.* $Act \overset{def}{=} Imp \cup \overline{Imp} \cup Exp^- \cup \overline{Exp^-} \cup \{\tau\}$;
- *A set $D \subseteq Q \times Act \times Q$ of discrete transitions. The action a is* enabled *in x if there exists a x' such that $(x, a, x') \in D$;*
- *A set $\mathcal{T}$ of* trajectories *for Var. Each $f \in \mathcal{T}$ is a function whose domain, $dom(f)$, is an initial subset of $\mathbb{R}_{\geq 0}$ and whose image set is a subset of Q, i.e., $f(t) \in Q$ for all $f \in \mathcal{T}$ and all $t \in dom(f)$. The following axioms must hold:*
  T1 *(Prefix closure)*
        If $f \in \mathcal{T}, dom(f') \subseteq dom(f)$, and $f'(t) = f(t)$ for all $t \in dom(f')$, then $f' \in \mathcal{T}$;
  T2 *(Suffix closure)*
        If $f \in \mathcal{T}$ and $t' \in dom(f)$, then $f'(t) \overset{def}{=} f(t + t')$ belongs to $\mathcal{T}$;

T3 *(Concatenation closure)*
Let $S = \{f_0, f_1, f_2, \ldots\}$ be a subset of $\mathcal{T}$ such that $dom(f_i)$ has the form $[0, t_{max,i}]$ and $f_i(t_{max,i}) = f_{i+1}(0)$ for all $i + 1 < |S|$. Then the trajectory:

$$(f_0 \frown f_1 \frown f_2 \frown \ldots)(t) \overset{def}{=} \begin{cases} f_0(t) & \text{if } t \in dom(f_0) \\ (f_1 \frown f_2 \frown \ldots)(t - t_{max,0}) & \text{otherwise} \end{cases}$$

belongs to $\mathcal{T}$.

The trajectories can be given in implicit form, for instance, as a differential system: the set $\mathcal{T}$ contains all the solutions of the differential system. In such cases, the computation of the trajectories themselves is not always trivial and may be not computable.

The semantics of hybrid automata can be given in terms of MLTS's by associating a hybrid automaton with an infinite MLTS whose nodes are the states of the automaton and edges have the form $\overset{t}{\to}_C$ or $\overset{e}{\to}_D$. The *continuous transition relation* $q \overset{t}{\to}_C q'$ holds if and only if there exists a $f \in \mathcal{T}$ such that $[0, t] \subseteq dom(f)$, $f(0) = q$ and $f(t) = q'$. The *discrete transition relation* $q \overset{a}{\to}_D q'$ holds if and only if $(q, a, q') \in D$. We may write $q \overset{a}{\to} q'$ to denote that either $q \overset{a}{\to}_C q'$ or $q \overset{a}{\to}_D q'$.

Definition 5 (Hybrid Automata - Semantics). *Let $\mathcal{A}$ be the hybrid automaton (Imp, Exp, Q, Θ, Imp, Exp^-, D, $\mathcal{T}$). The MLTS $L(\mathcal{A})$ associated to $\mathcal{A}$ is the tuple $(Q, \Theta, Imp \cup Exp, R)$, where $Exp \overset{def}{=} Exp^- \cup \mathbb{R}_{\geq 0}$ and $R \overset{def}{=} \{(q, \alpha, q') \mid q \overset{\alpha}{\to} q' \wedge \alpha \in (\mathbb{R}_{\geq 0} \cup Act)\}$.*

In the above definition, we consider all the actions associated to the continuous transitions, i.e., the positive real numbers labeling the transition relation $\to_C$, as exposed actions. This is an arbitrary choice. The imposed interactions with the environment are represented only through some discrete transition labels, but they could indirectly influence both the continuous and the discrete exposed behaviors. For those who are more familiar with the definition of hybrid automata given in [1,24], this means that the imposed interactions can cause a change of location in the automaton and hence a possible change in the differential laws regulating the continuous evolution.

Example 1 (Simple thermostat model). Let us model a simple thermostat by using a hybrid automaton. The discrete variable *mode* represents the state of the heater (i.e., *mode* = 1 means "heater on" and *mode* = 0 "heater off"), while the variable x_T is associated to the temperature. Whenever the temperature reaches $15\,°C$, the thermostat activates the heater (exposed action *switchOn*), while, if the temperature rises up to $20\,°C$, the heater is turned off (exposed action *switchOff*). The users can switch on and off the heater independently from the thermostat status by using the imposed actions *forceOn* and *forceOff*, respectively.

The two constants k_r and k_h are the dispersion and heating coefficients, respectively, while the variable X represents the room temperature. Initially, the heater is switched off and the temperature is $17\,°C$.

In the context of hybrid automata, (bi)simulation reductions [2,30], series of abstractions [44], piecewise linear approximations [19,20] have been proposed to abstract the infinite MLTS of an hybrid automaton into finite ones.

An hybrid automaton whose set of trajectories is empty are said to be a *(hybrid) discrete automaton*.

Table 1. The discrete components of the hybrid automaton modeling a thermostat

Meaning	Actions
Turn off the heater	*switchOff*
Turn on the heater	*switchOn*
Force off the heater	*forceOff*
Force on the heater	*forceOn*

(a) Actions

$Imp \setminus Exp$	**Preconditions**	**Action**	**Effects**
Exp	$(mode = 1) \wedge (x_T \geq 20)$	*switchOff*	$mode \leftarrow 0$
Exp	$(mode = 0) \wedge (x_T \leq 15)$	*switchOn*	$mode \leftarrow 1$
Imp	$mode = 1$	*forceOff*	$mode \leftarrow 0$
Imp	$mode = 0$	*forceOn*	$mode \leftarrow 1$

(b) Transitions

Example 2 (Discrete automaton). Let us consider a discrete automaton D modeling the rebound heights of *bouncing ball* whose collisions are inelastic. The automaton is equipped with two variables, x and *mode*: the former represents ball's elevation, while the latter is set to 1 as long as the ball is bouncing and become 0 as soon as the ball does not bounce anymore.

Table 3 reports both the actions and the transitions of D. The symbol γ represents the ball coefficient of restitution. The trajectories of this automaton are constant functions and, thus, D is a discrete automaton. By imposing as starting height $x = 10$ and as coefficient of restitution $\gamma = 0.86$, the bounce peaks decrease at each iteration.

2   Interactions with the Environment

We use hybrid automata as modeling language for both the system of interest and the environment. The environment can interact with the system and change its behavior. For instance, a pathogen agent in the environment could activate the immune response of the system under analysis. In this paper, we limit our analysis to environments represented by discrete automata and, in this section, we formalize how such discrete automata interact with generic hybrid automata.

Given a hybrid automaton $\mathcal{A}$ and a discrete automaton $\mathcal{D}$ that have different variables and share the same set of actions, we can define the hybrid automaton $\mathcal{A} \| \mathcal{D}$ that is the composition of $\mathcal{A}$ and $\mathcal{D}$.

Definition 6 (Composition of automata). *Let $\mathcal{A}$ be the hybrid automata $(Imp_A, Exp_A, Q_A, \Theta_A, Imp, Exp^-, D_A, \mathcal{T}_A)$ and let $\mathcal{D}$ be and the discrete automata $(Imp_D, Exp_D, Q_D, \Theta_D, Imp, Exp^-, D_D, \emptyset)$, where $Imp_D \cap Imp_A = \emptyset$ and $Exp_D \cap Exp_A = \emptyset$. The composition of $\mathcal{D}$ on $\mathcal{A}$, denoted by $\mathcal{A} \| \mathcal{D}$, is the hybrid automaton $(Imp, Exp, Q, \Theta, Imp, Exp^-, D, \mathcal{T})$ where:*

- *$Imp = Imp_A \cup Imp_D$ and $Exp = Exp_A \cup Exp_D$;*

Table 2. The continuous components of the hybrid automaton modeling a thermostat

Trajectories
$\dot{x_T} = k_h * mode - k_r * x_T$
$\dot{mode} = 0$

(a) Trajectories

Meaning	Variables
Temperature	$x_T : \mathbb{R} \leftarrow 17,$
Heater state	$mode : \{0, 1\} \leftarrow 0$

(b) Internal Variables

Table 3. Actions and transitions of a discrete automaton representing a bouncing ball

Meaning	Actions		$Imp \setminus Exp$	Preconditions	Action	Effects
The ball bounced	*bounce*		Exp	$(mode = 1 \wedge x > 0)$	*bounce*	$x \leftarrow \gamma * x$
The ball stops to move	*stop*		Exp	$(mode = 1 \wedge x = 0)$	*stop*	$mode \leftarrow 0$

(a) Actions	(b) Transitions

- $Q = Q_A \times Q_D$ and $\Theta = \Theta_A \times \Theta_D$;
- $((u, u'), \alpha, (v, v')) \in D$ iff either
  * $u \xrightarrow{\alpha}_D v$ in D_A and $u' = v' \in Q_D$ or
  * $u = v \in Q_A$ and $u' \xrightarrow{\alpha}_D v'$ in D_D or
  * $\alpha = \tau$ and $\exists \beta \in Imp \cup Exp^-$ s.t. $u \xrightarrow{\beta}_D v$ in D_A and $u' \xrightarrow{\beta}_D v'$ in D_D or
  * $\alpha = \tau$ and $\exists \beta \in Imp \cup Exp^-$ s.t. $u \xrightarrow{\overline{\beta}}_D v$ in D_A and $u' \xrightarrow{\overline{\beta}}_D v'$ in D_D;
- $f \in \mathcal{T}$ iff $\exists f_A \in \mathcal{T}_A$ s.t. $f(t) = (f_A(t), u')$ for all $t \in dom(f)$ and all $u' \in Q_D$.

The semantics of $\mathcal{A} \| \mathcal{D}$ is fully specified by $L(\mathcal{A} \| \mathcal{D})$. However, it is worth to investigate the relation between the semantics of $\mathcal{A}$, $\mathcal{D}$, and $\mathcal{A} \| \mathcal{D}$. In order to achieve this goal, we need to formalize the evolution of the MLTS's representing two interacting systems that run independently and synchronize whenever it is possible. Such composed evolutions can be modeled through the MLTS defined as follows.

Definition 7 (Composition $\|$). *Let* $T^1 = (V^1, V_I^1, Imp \cup Exp, E^1)$ *and* $T^2 = (V^2, V_I^2, Imp \cup Exp, E^2)$ *be two MLTS's over a common set of input edge labels. The composition of* T^1 *and* T^2 *is the MLTS* $T^1 \| T^2 = (V^{12}, V_I^{12}, Imp \cup Exp, E^{12})$, *where:*

- $V^{12} = V^1 \times V^2$ *and* $V_I^{12} = V_I^1 \times V_I^2$;
- $(u_1, u_2) \xrightarrow{\alpha} (v_1, v_2)$ *is in* E^{12} *if and only if one of the following conditions hold:*
  * $u_1 \xrightarrow{\alpha} v_1$ *and* $u_2 = v_2$;
  * $u_2 \xrightarrow{\alpha} v_2$ *and* $u_1 = v_1$;
  * $\alpha = \tau$ *and there exists* $\beta \in Imp \cup Exp$ *such that* $u_1 \xrightarrow{\beta} v_1$ *and* $u_2 \xrightarrow{\overline{\beta}} v_2$ *or vice-versa.*

For those who are familiar with process algebras, again this coincides with the parallel composition of CCS. If T_1 is the MLTS representing the CCS process E_1 and T_2 is the MLTS representing the CCS process E_2, then $T_1 \| T_2$ represents the CCS process $E_1 | E_2$.

The following theorem links the semantics of $\mathcal{A}$ and $\mathcal{D}$ with that of $\mathcal{A} \| \mathcal{D}$.

Theorem 1. *Let* $\mathcal{A}$ *be the hybrid automaton* $(Imp_A, Exp_A, Q_A, \Theta_A, Imp, Exp^-, D_A, \mathcal{T}_A)$ *and let* $\mathcal{D}$ *be the discrete automaton* $(Imp_D, Exp_D, Q_D, \Theta_D, Imp, Exp^-, D_D, \emptyset)$, *where* $Imp_D \cap Imp_A = \emptyset$ *and* $Exp_D \cap Exp_A = \emptyset$. *The MLTS* $L(\mathcal{A} \| \mathcal{D})$ *is equal to the composition of the two MLTS's* $L(\mathcal{A})$ *and* $L(\mathcal{D})$, *i.e.,* $L(\mathcal{A} \| \mathcal{D}) = L(\mathcal{A}) \| L(\mathcal{D})$

Proof. By Definition 5, $L(\mathcal{A})$ is the MLTS $(Q_A, \Theta_A, Imp \cup Exp, R_A)$, where $Exp = Exp^- \cup \mathbb{R}_{\geq 0}$ and $R_A \stackrel{def}{=} \{(q, \alpha, q') | q \xrightarrow{\alpha} q' \wedge \alpha \in (\mathbb{R}_{\geq 0} \cup Act)\}$. Since $\mathcal{D}$ has no trajectories, $L(\mathcal{D})$ is the MLTS $(Q_D, \Theta_D, Imp \cup Exp, R_D)$, where $Exp = Exp^- \cup \mathbb{R}_{\geq 0}$ and

$R_D \stackrel{def}{=} \{(q, \alpha, q') \mid q \stackrel{\alpha}{\to}_D q' \wedge \alpha \in Act\}$. Thus, by Definition 7, $L(\mathcal{A}) \| L(\mathcal{D})$ is the MLTS $(Q_A \times Q_D, \Theta_A \times \Theta_D, Imp \cup Exp, R)$, where $(u, u') \stackrel{\alpha}{\to} (v, v')$ is in R if and only if one of the following conditions hold (a) $u \stackrel{\alpha}{\to} v$ in R_A and $u' = v'$, (b) $u' \stackrel{\alpha}{\to} v'$ in R_D and $u = v$, (c) $\alpha = \tau$ and there exists $\beta \in Imp \cup Exp$ such that $u \stackrel{\beta}{\to} v$ in R_A and $u' \stackrel{\bar{\beta}}{\to} v'$ in R_D or (d) vice-versa.

By Definitions 6, $\mathcal{A} \| \mathcal{D}$ is the hybrid automaton $(Imp_A \cup Imp_D, Exp_A \cup Exp_D, Q_A \times Q_D, \Theta_A \times \Theta_D, Imp, Exp^-, D, \mathcal{T})$ where:

- $((u, u'), \alpha, (v, v')) \in D$ if and only if either
  * $u \stackrel{\alpha}{\to}_D v$ in D_A and $u' = v' \in Q_D$ or
  * $u = v \in Q_A$ and $u' \stackrel{\alpha}{\to}_D v'$ in D_D or
  * $\alpha = \tau$ and $\exists \beta \in Imp \cup Exp^-$ s.t. $u \stackrel{\beta}{\to}_D v$ in D_A and $u' \stackrel{\beta}{\to}_D v'$ in D_D or
  * $\alpha = \tau$ and $\exists \beta \in Imp \cup Exp^-$ s.t. $u \stackrel{\bar{\beta}}{\to}_D v$ in D_A and $u' \stackrel{\bar{\beta}}{\to}_D v'$ in D_D;
- $f \in \mathcal{T}$ iff $\exists f_A \in \mathcal{T}_A$ s.t. $f(t) = (f_A(t), p)$ for all $t \in dom(f)$ and all $p \in Q_D$.

Thus, by Definition 5, $L(\mathcal{A} \| \mathcal{D})$ is the MLTS $(Q, \Theta, Imp \cup Exp, R')$ where $Q = Q_A \times Q_D$, $\Theta = \Theta_A \cup \Theta_D$, and $R' = \{((f(0), p), t, (f(t), p)) \mid f \in \mathcal{T}_A \wedge t \in dom(f) \wedge p \in Q_D\} \cup D$. It is easy to see that $R' = R$ and, hence, the thesis holds.   □

3   Robustness as Unwinding

As observed in [29], "...robustness is one of the fundamental characteristics of bio-logical systems ... Nevertheless, a mathematical foundation that provides a unified perspective on robustness is yet to be established". In [28], robust systems maintain their functions against (internal and external) perturbations. Hence, any framework for the analysis of robustness should support the definition of both functions to be preserved and admissible perturbations. Robustness is different from stability: robust systems can exploit instability or even evolve through new steady states in order to preserve their functions against perturbations.

Example 3 (Pathogen agents). Let us consider the case of a system infected by a pathogen agent. The pathogen stimulates the immune system and would probably affect some organs, i.e., the system is not stable. Hence, the immune system and all the organs directly attacked by the pathogen will exhibit a behavior which could be very different from the standard one (at least in the acute phase). However, if the system is robust against the pathogen, the behaviors of critical organs (e.g., heart, lung, brain) should not be dramatically affected.

A formalization of "how much" a system is *robust* with respect to a pathogen is a fundamental question in medicine both in the diagnosis process (to avoid expensive/invasive exams) and in the therapy phase.

On the other hand, if a system is *stable* with respect to a pathogen, this means that the pathogen has almost no effects on the system. Hence, probably such pathogen has low medical interest: not even the patient will notice that his immune system is interacting with the pathogen.

We propose to informally express robustness as follows. A biological system S which can either interact with the environment through imposed actions or perform exposed (internal) actions is said to be *robust* if the imposed behaviors do not influence the exposed ones. This is nothing but a biological reformulation of the notion of non-interference introduced in [21] in the area of information flow security.

Example 4 (Pathogen agents – cont). Let us consider again the system infected by a pathogen agent. The pathogen agent interacts with the system through imposed actions. In particular it interacts with the immune system and with all the organs it directly attacks. On the other hand, the organs which are not directly attacked (e.g., heart, lung, kidney, brain) will perform exposed actions. If we do not want them to be dramatically affected their behavior should be preserved. In other words, some changes in the immune system (imposed actions) could slightly modify the usual behavior of the "exposed" organs. To be acceptable such modified evolutions should be behaviorally equivalent to the usual ones. Notice that behaviorally equivalent evolutions could have many differences depending on how coarse is the equivalence.

In the field of information flow security, non-interference has been formalized through (generalized) unwinding conditions (see, e.g., [40,5,4]). Here we introduce unwinding conditions, following the approach of [5,4], with the aim of formalizing robustness.

Definition 8 (Robustness through Unwinding). *Let* $T = (V, V_I, Imp \cup \mathcal{E}xp, E)$ *be a MLTS representing a biological system,* $=^{\mathcal{E}xp}, \sim^{\mathcal{E}xp} \subseteq V \times V$ *be two equivalence relations, and* $\dashrightarrow \subseteq V \times V$ *be a transition relation. We say that* T *is unwinding robust, denoted as* $T \in \mathcal{W}(= \mathcal{E}xp, \sim^{\mathcal{E}xp}, \dashrightarrow)$, *if for each* $\iota \in Imp \cup \overline{Imp}$ *for each* $u \in V$ *the following condition holds:*

$$u \xrightarrow{\iota} u' \text{ implies } \forall v(u =^{\mathcal{E}xp} v \text{ implies } \exists v'(v \dashrightarrow v' \wedge u' \sim^{\mathcal{E}xp} v'))$$

Unwinding conditions can be introduced over any modeling language whose operational semantics is defined through MLTS's. Moreover, they can be instantiated in an infinite number of ways choosing the exposed behavioral equivalences $=^{\mathcal{E}xp}$ and $\sim^{\mathcal{E}xp}$ and the transition relation $\dashrightarrow$. The exposed behavioral equivalences $= \mathcal{E}xp$ and $\sim^{\mathcal{E}xp}$ establish which variations in the behavior of the system are considered acceptable. The transition relation $\dashrightarrow$ has a less intuitive meaning. It represents a possible *delay* or more in general a *temporal misalignment* between the behavior of the system running in normal conditions and the system after an *imposed* change.

Going back to Kitano's requirements for a robustness framework [28], in our notion of robustness based on unwinding conditions, the functions of a system are mainly defined through the *exposed observational equivalences* $=^{\mathcal{E}xp}$ *and* $\sim^{\mathcal{E}xp}$, while the perturbations are modeled through *imposed actions*.

Recently many authors have proposed to model robustness in the biological setting as "properties preservation". Temporal Logics can be used as specification languages for expressing the properties of interest. In such context robustness has been defined as a way to "measure the distance" (or in other terms a "degree of satisfaction") of a set of traces from a given specification (see, e.g., [15,14,39,13,6,3]). In [9] we show that unwinding conditions can be instantiated to obtain a robustness notion equivalent to the one defined in [15].

At this point there is still a missing connection. In Section 2 we formalized the interactions between a system and the environment through a composition of systems. In this section we introduced unwinding conditions in which the interactions with the environment are represented by imposed transitions: each time an imposed transition is performed (e.g., $u \xrightarrow{l} u'$) some conditions hold (e.g., $\forall v(u =^{\mathcal{E}xp} v$ implies $\exists v'(v \dashrightarrow v' \wedge u' \sim^{\mathcal{E}xp} v')))$. Which are the relationships between these two apparently different notions of interaction? In the next section, we show an instance of the unwinding framework over hybrid automata in which systems satisfying the unwinding condition can safely interact with any environment performing imposed actions. This is not true in general for any unwinding condition. As a matter of facts, unwinding conditions can be defined also over languages which do not admit system composition.

4   Unwindings over Hybrid Automata

We now introduce an unwinding condition over hybrid automata, i.e., over the MLTS's defining the semantics of hybrid automata.

Definition 9 (Exposed Weak Bisimulation $\approx^{\mathcal{E}xp}$). *Given a MLTS* $T = (V, V_I, Imp \cup \mathcal{E}xp, E)$, *an* exposed weak bisimulation *over T is a symmetric relation $R \subseteq V \times V$ such that for each $(u, v) \in R$ the following conditions hold:*

- *$u \in V_I$ if and only if $v \in V_I$;*
- *for each $\epsilon \in \mathcal{E}xp \cup \overline{\mathcal{E}xp} \cup \{\tau\}$ if $u \xrightarrow{\epsilon} u'$, then $v \xRightarrow{\hat{\epsilon}} v'$ and $(u', v') \in R$.*

Two nodes $u, v \in V$ are said to be exposed weakly bisimilar, *denoted as $u \approx^{\mathcal{E}xp} v$, if there exists a weak bisimulation R such that $(u, v) \in R$.*

When T_1 and T_2 are defined over the same sets $Imp \cup \mathcal{E}xp$ we will use the notation $T_1 \approx^{\mathcal{E}xp} T_2$ to denote that there exists an exposed weak bisimulation R such that for each node v_1 of T_1 there exists at least one node v_2 of T_2 such that $(v_1, v_2) \in R$ and vice-versa. Moreover, if $(v_1, v_2) \in R$, then v_1 is an initial node in T_1 if and only if v_2 is an initial node in T_2.

Definition 10 (Weak Unwinding $\mathcal{W}(Id, \approx^{\mathcal{E}xp}, (\xrightarrow{\tau})^*)$). *We say that a MLTS T satisfies the* weak unwinding *if $T \in \mathcal{W}(Id, \approx^{\mathcal{E}xp}, (\xrightarrow{\tau})^*)$, where Id is the identity relation.*

A hybrid automaton $\mathcal{H}$ satisfies the weak unwinding *if the MLTS $L(\mathcal{H})$ does.*

In the above definition if a system satisfies the weak unwinding, then whenever a node u performs an imposed transition and it reaches a node u', there must exist a sequence of silent transitions which leads to a node exposed weakly bisimilar to u'. Weak unwinding over MLTS's is just a transposition in the biological setting of the definition of the security property P_BNDC over the process algebra SPA, a multilevel version of CCS (see, e.g., [17,5]).

We now prove that whenever a hybrid automaton satisfies the weak unwinding it can safely interact with any discrete environment that performs only imposed transitions. Notice that it is meaningless to ask what happens when a system interacts with an environment performing also exposed transitions. Of course, in such a case, the environment can directly change the exposed behavior and there is no chance to preserve it.

Definition 11 (Imposed Environments). *Let* $\mathcal{I} = (V, V_I, Imp \cup \mathcal{E}xp, E)$ *be a MLTS we say that* $\mathcal{I}$ *is an* imposed MLTS *if in* $\mathcal{I}$ *all the edges have labels which range in* $Imp \cup \overline{Imp} \cup \{\tau\}$ *and if* $u \xrightarrow{\tau} u'$, *then* $u \in V_i$ *iff* $u' \in V_i$.

Let $\mathcal{D}$ *be a discrete hybrid automaton. We say that* $\mathcal{D}$ *is an* imposed environment *if the MLTS* $L(\mathcal{D})$ *is an imposed MLTS.*

We are now ready to establish the connection between weak unwinding and interactions with imposed environments.

Theorem 2. *Let* $T = (V, V_I, Imp \cup \mathcal{E}xp, E)$ *be a MLTS. If* $T \in \mathcal{W}(Id, \approx^{\mathcal{E}xp}, (\xrightarrow{\tau})^*)$, *then for each imposed MLTS* $\mathcal{I}$ *it holds that* $T\|\mathcal{I} \approx^{\mathcal{E}xp} T$.

Proof. Let $\mathcal{S} = \{((u, i), u') \mid u \approx^{\mathcal{E}xp} u', i$ is a node of $\mathcal{I}$, and $u \in V_i$ iff is initial in $\mathcal{I}\}$. We prove that $\mathcal{S} \cup \mathcal{S}^{-1}$ is an exposed weak bisimulation between $T\|\mathcal{I}$ and T.

It is immediate to see that for each element (u, i) of $T\|\mathcal{I}$ there is at least $((u, i), u)$ in $\mathcal{S}$ and for each element u of T there is at least $(u, (u, i)) \in \mathcal{S}^{-1}$.

Let $((u, i), u') \in \mathcal{S}$. If $(u, i) \xrightarrow{\epsilon} (v, j)$ with $\epsilon \in \mathcal{E}xp \cup \overline{\mathcal{E}xp} \cup \{\tau\}$, then we have to prove that u' is able to simulate such transition. Three cases have to be considered. If $(u, i) \xrightarrow{\epsilon} (v, i)$ since $u \xrightarrow{\epsilon} v$, then since $u \approx^{\mathcal{E}xp} u'$ it holds that $u' \xrightarrow{\hat{\epsilon}} v'$ with $v \approx^{\mathcal{E}xp} v'$ and hence $((v, i), v') \in \mathcal{S}$. If $(u, i) \xrightarrow{\epsilon} (u, j)$ since $i \xrightarrow{\epsilon} j$, then since $\mathcal{I}$ is an imposed MLTS, it holds that $\epsilon = \tau$. Moreover, $u'(\xrightarrow{\tau})^0 u'$ and $((u, j), u') \in \mathcal{S}$. If $(u, i) \xrightarrow{\tau} (v, j)$ since $u \xrightarrow{\iota} v$ and $i \xrightarrow{\bar{\iota}} j$ (or vice-versa) with $\iota \in Imp$, then since $T \in \mathcal{W}(Id, \approx^{\mathcal{E}xp}, (\xrightarrow{\tau})^*)$ it holds that $u(\xrightarrow{\tau})^* v'$ with $v \approx^{\mathcal{E}xp} v'$. Hence, since $u \approx^{\mathcal{E}xp} u'$ it holds that $u(\xrightarrow{\tau})^* v''$ with $v' \approx^{\mathcal{E}xp} v''$. So, $v \approx^{\mathcal{E}xp} v''$ and $((v, j), v'') \in \mathcal{S}$.

Let $(u', (u, i)) \in \mathcal{S}^{-1}$. If $u' \xrightarrow{\epsilon} v'$ with $\epsilon \in \mathcal{E}xp \cup \overline{\mathcal{E}xp} \cup \{\tau\}$, then we have to prove that (u, i) is able to simulate such transition. Since $u \approx^{\mathcal{E}xp} u'$ we have that $u \xrightarrow{\hat{\epsilon}} v$ with $v \approx^{\mathcal{E}xp} v'$. Hence, $(u, i) \xrightarrow{\hat{\epsilon}} (v, i)$ and $(v', (v, i)) \in \mathcal{S}^{-1}$. $\square$

Corollary 1. *Let* $\mathcal{H}$ *be a hybrid automaton which satisfies the weak unwinding. For any imposed environment* $\mathcal{D}$ *it holds that* $\mathcal{H}\|\mathcal{D} \approx^{\mathcal{E}xp} \mathcal{H}$.

For readers familiar with security properties, this result is similar to the one which states that *P_BNDC* implies *BNDC*. The main difference is that we are using hybrid automata as modeling language instead of a CCS-like process algebra. As a consequence we are working on bisimulations defined over directed (possibly not connected, infinite and dense) labeled graphs with sets of initial states. All the MLTS's which can be defined through CCS can be defined through hybrid automata, while the opposite does not hold.

The converse of the above corollary does not hold. However, if we allow to change the set of initial states, then we get the following characterization of weak unwinding.

Theorem 3. *Let* $T = (V, V_I, Imp \cup \mathcal{E}xp, E)$ *be a MLTS.* $T \in \mathcal{W}(Id, \approx^{\mathcal{E}xp}, (\xrightarrow{\tau})^*)$ *if and only if for each* T' *obtained from* T *by changing the set* V_I *and for each imposed MLTS* $\mathcal{I}$ *it holds that* $T'\|\mathcal{I} \approx^{\mathcal{E}xp} T'$.

Proof. If $T \in \mathcal{W}(Id, \approx^{Exp}, (\xrightarrow{\tau})^*)$, then the relation $\mathcal{S} \cup \mathcal{S}^{-1}$ used in the proof of Theorem 2 works for any T' obtained from T by changing the set V_I since it contains all the pairs of the form $(v, (v, i))$ and $(v, (v, i))$.

On the other hand, let us assume that for each T' obtained from T by changing V_I it holds that $T\|I \approx^{Exp} T'$. Let u be a node of T. We have to prove that if $u \xrightarrow{\iota} v$ with $\iota \in Imp$, then $u(\xrightarrow{\tau})^* v'$ with $v \approx^{Exp} v'$. Let us consider T' such that $V_I = \{u\}$ and I be the MLTS having two nodes i and j with i initial node and $i \xrightarrow{\iota} j$. We have that in $T'\|I$ the node (u, i) is the only initial one and $(u, i) \xrightarrow{\tau} (v, j)$. Hence, since $T'\|I \approx^{Exp} T'$, it holds that $u(\xrightarrow{\tau})^* v'$ such that $(v, j) \approx^{Exp} v'$. Since, j has no outgoing edges we also get $(v, j) \approx^{Exp} v$. Hence $v \approx^{Exp} v'$.   □

Corollary 2. *Let $\mathcal{H}$ be a hybrid automaton. $\mathcal{H}$ satisfies the weak unwinding if and only if for any $\mathcal{H}'$ obtained from $\mathcal{H}$ by changing the set Θ and for any imposed environment $\mathcal{D}$ it holds that $\mathcal{H}'\|\mathcal{D} \approx^{Exp} \mathcal{H}'$.*

Again as in the case of the security property *P_BNDC*, this result states that an automaton satisfies the weak unwinding if and only if it can safely interact with any hostile, possibly dynamically, changing environment. If the environment can only execute imposed actions, then it cannot dramatically modify the exposed behavior of the system.

We now introduce two further unwinding conditions to give to the reader an idea of the flexibility of the framework. We consider two nodes of the MLTS to be "equal" if the have the same values on the exposed variables . We consider them to be "equivalent" if they are "equal" and using only exposed transitions they generate the same sequences of exposed variables values. The two unwindings only differ on the instantiation of the transition relation $\dashrightarrow$.

Definition 12 ($\doteq^{Exp}$, $\preccurlyeq^{Exp}$, $\Rightarrow^{Exp}$ **and** $\rightarrow^{>t}$). *Let $\mathcal{H}$ be a hybrid automaton and $L(\mathcal{H}) = (Q, \Theta, Imp \cup Exp, R)$ be the MLTS associated to $\mathcal{H}$. The relation $\doteq^{Exp} \subseteq Q \times Q$ is defined as follows: $u \doteq^{Exp} v$ if and only if the exposed variables have the same values on u and v. A bi-exposed weak bisimulation over $L(\mathcal{H})$ is a symmetric binary relation R over $L(\mathcal{H})$ such that for each $(u, v) \in R$:*

- $u \doteq^{Exp} v$;
- *for each $\epsilon \in Exp \cup \overline{Exp} \cup \{\tau\}$ if $u \xrightarrow{\epsilon} u'$, then $v \xRightarrow{\hat{\epsilon}} v'$ and $(u', v') \in R$.*

Two states u and v are bi-exposed weakly bisimilar, *denoted by $u \preccurlyeq^{Exp} v$ if there exists a bi-exposed weak bisimulation R such that $(u, v) \in R$.*

The relations $\Rightarrow^{Exp} \subseteq Q \times Q$ and $\rightarrow^{>t} \subseteq Q \times Q$, where $t \in \mathbb{R}_{\geq 0}$, are defined as follows: $u \Rightarrow^{Exp} v$ if and only if there exists $\epsilon \in Exp \cup \overline{Exp} \cup \{\tau\}$ such that $u \xRightarrow{\hat{\epsilon}} v$ and $u \rightarrow^{>t} v$ if and only if there exists $t' > t$ such that $u \xrightarrow{t'} v$.

Definition 13 (Bi-Exposed Unwinding and Delayed Unwinding). *$\mathcal{H}$ satisfies the bi-exposed unwinding if $L(\mathcal{H}) \in \mathcal{W}(\doteq^{Exp}, \preccurlyeq^{Exp}, \Rightarrow^{Exp})$. $\mathcal{H}$ satisfies the t-delayed bi-exposed unwinding if $L(\mathcal{H}) \in \mathcal{W}(\doteq^{Exp}, \preccurlyeq^{Exp}, \rightarrow^{>t})$.*

Since continuous transitions are exposed, if a system satisfies the t-delayed bi-exposed unwinding, then it also satisfies the bi-exposed one.

In the next section we apply weak and bi-exposed unwindings in the analysis of influenza models, proving that, as one could expect, influenza is not robust with respect to antiviral and interferon treatments. Moreover, t-delayed bi-exposed unwinding allows to show that also if we refer to a single type of treatment (e.g., antiviral) the result strongly depends on the time at which the treatment is imposed.

5 Influenza Kinetics Analysis

Influenza is an infectious disease caused by a family of RNA viruses known as influenza viruses. Its symptoms include fever, weakness, and coughing and, in the most acute form, it can bring a severe threat to the respiratory system. Worldwide, up to 500,000 deaths per year are due to the complications of the seasonal influenza virus [34] and the infamous *Spanish flu* of the 1918 infected 500 million people leading to the death of 20 to 100 million of them [26,43].

Understanding the dynamics of infection plays a crucial role in avoiding or, at least, controlling possible influenza pandemics. Many models have been suggested so far to achieve this goal [23,42].

Handel *et al.* deal with the effects of the most effective drugs against influenza, the *Neuraminidase inhibitors* (NI), and take into account the rise of virus strains resistant to this antiviral [23]. The model distinguishes the load of viruses that are NI-sensitive (V_s) from that of viruses that are NI-resistant (V_r). Uninfected cells (U) are infected by either NI-resistant virus or NI-sensitive virus at a rate proportional to the correspondent virus load and became NI-resistant infected (I_r) or NI-sensitive infected (I_s), respectively. Both NI-resistant and NI-sensitive infected cells increase virus load of the respective strain. Moreover, due to natural mutations, a fraction of the viruses released by NI-sensitive infected cells belongs to the NI-resistant strain. Whatever is the strain of the viruses produced by NI-sensitive infected cells, NI represses their release at a rate that is proportional to the efficiency of the antiviral itself. A natural immune response (X) restrains the increase in viral load in all the strains too. As it occurs also to the virus strains, both the kinds of infected cells share the same decay rate.

On the contrary, Saenz *et al.* consider the interactions between viral agents and immune system and describe the antiviral response modulated by the type I interferon (IFN-α/β) which is triggered by infection [42]. In their model, uninfected cells (U) are infected at a rate proportional to the virus load (V). Newly infected cells spend some time in an eclipse phase (E_1) and then they move to a state (I), having a limited span life, in which they increase the virus load. Infective cells produce IFN (F) which is able to bring uninfected cells into a prerefractory state (W) and, possibly, in a refractory cells (R) at a rate proportional to F itself. Whenever cells in the prerefractory state are infected, they move to an eclipse phase (E_2), in which they release IFN, and, eventually, become infective (I).

IFNs have been long used as a treatment for various autoimmune, viral, and tumor diseases [38,22,25]. So, one may wonder if IFNs can be used as antiviral drugs and, in particular, whether IFNs and NIs are equivalent or not with respect to the virus load,

Table 4. Variables of the hybrid automaton modeling the influenza kinetics

Meaning	Variables
Uninfected cells	$U : \mathbb{R}_{\geq 0} \leftarrow 4.0 * 10^8$
Cells infected by virus sensitive to NI in eclipse phase	$E_{1,s} : \mathbb{R}_{\geq 0} \leftarrow 0.0$
Cells infected by virus resistant to NI in eclipse phase	$E_{1,r} : \mathbb{R}_{\geq 0} \leftarrow 0.0$
Prerefractory cells infected by virus sensitive to NI in eclipse phase	$E_{2,s} : \mathbb{R}_{\geq 0} \leftarrow 0.0$
Prerefractory cells infected by virus resistant to NI in eclipse phase	$E_{2,r} : \mathbb{R}_{\geq 0} \leftarrow 0.0$
Cells in prerefractory state	$W : \mathbb{R}_{\geq 0} \leftarrow 0.0$
Refractory cells	$R : \mathbb{R}_{\geq 0} \leftarrow 0.0$
Cells infected by virus sensitive to NI	$I_s : \mathbb{R}_{\geq 0} \leftarrow 0.0$
Cells infected by virus resistant to NI	$I_r : \mathbb{R}_{\geq 0} \leftarrow 0.0$
Virus sensitive to NI	$V_s : \mathbb{R}_{\geq 0} \leftarrow 7.7 * 10^{-3}$
Virus resistant to NI	$V_r : \mathbb{R}_{\geq 0} \leftarrow 0.0$
Interferon (IFN-α/β)	$F : \mathbb{R}_{\geq 0} \leftarrow 0.0$
Immune response	$X : \mathbb{R}_{\geq 0} \leftarrow 3.4 * 10^{-1}$
Elapsed time	$T : \mathbb{R}_{\geq 0} \leftarrow 0.0$
Antiviral injection mode ($A = 0$ not injecting NI, $A = 1$ injecting NI)	$A : \{0, 1\} \leftarrow 0$
Model mode ($M = 0$ treatment to be selected, $M = 1$ NI-based treatment, $M = 2$ IFN-based treatment, $M = 3$ treatment concluded)	$M : [0, 3] \leftarrow 0$

i.e., if IFNs can be used in place of NIs, and vice-versa, in influenza treatment. Notice that, whatever the answer is, the NI-based therapies will be still preferable to the IFN-based ones in normal condition since the latter exhibit many serious side effects in humans (see, e.g., [45]) and they are less cost-effective than the former. However, above questions maintain some relevance in the case of a pandemic produced by a viral strain that is resistant to the antivirals.

We developed a model that takes into account the effects of both IFN and NI on the virus load. As done in [23], we admit the existence of a NI-resistant strain and, analogously to [42], we represent the antiviral response due to IFN. The trajectories, the actions, and the transitions of our model are reported in Table 6. The values of their parameters are dependent on the virus strain and on the host species; we focused on human hosts infected with influenza A/Texas/91 (H1N1) and we fit our model on the data produced by the IR kinetic model described in [23] by minimizing the cost function:

$$SSE \overset{\text{def}}{=} \sum_i \left(\frac{\log V_s(i) - \log \hat{V}_s(i)}{\log \max_i \hat{V}_s(i)} \right)^2 + \sum_i \left(\frac{\log V_r(i) - \log \hat{V}_r(i)}{\log \max_i \hat{V}_r(i)} \right)^2 + (D - \hat{D})^2,$$

where V_s, V_r, and D are the number of NI-sensitive virus, the number of NI-resistant virus, the total death cell estimated by our model, respectively, while $\hat{V}_s$, $\hat{V}_r$, and $\hat{D}$ are the same quantities evaluated by the IR model suggested in [23].

Table 5. Trajectory parameters of the hybrid automaton modeling the influenza kinetics

Symbol	Meaning	Value	Source
k_i	Eclipse phase period	2 Days^{-1}	[27,42]
a	Prerefactory period	4 Days^{-1}	[12,42]
μ	Mutation rate	10^{-5}	[23]
σ	Fitness cost of resistance	0.1	[23]
n	IFN-reduced production	1	[42]
m	IFN-reduced infectivity	1	[42]
β	Virus infectivity	$1.18368 * 10^{-1}$	Fitted
p	Virus production per cell	$5.96623 * 10^{-4}$	Fitted
q	IFN induction per cell	$5.18653 * 10^{-11}$	Fitted
ψ	IFN efficiency	3.68109	Fitted
d_I	Infected cell death rate	$1.23661 * 10^{-1}$	Fitted
d_V	Virus death rate	$0.80710 * 10^{-1}$	Fitted
d_F	IFN clearance rate	1.80413	Fitted
r	Immune response growth rate	1.14835	Fitted
ϵ_{NI}	Antiviral efficacy	0.97647	Fitted
f	IFN released per injection	6	Arbitrary

Table 6. Trajectories, actions, and transitions of the influenza automaton

Trajectories

$$\dot{U} = -\beta(V_r + V_s)U - \phi FU$$
$$\dot{E}_{1,j} = \beta V_j U - k_1 E_{1,j}$$
$$\dot{E}_{2,j} = m\beta V_j W - k_2 E_{2,j}$$
$$\dot{W} = \phi FU - m\beta(V_r + V_s)W - aW$$
$$\dot{R} = aW$$
$$\dot{I}_j = k_1 E_{1,j} + k_2 E_{2,j} - d_I I_j$$
$$\dot{V}_s = (1 - A\epsilon_{NI})(1 - \mu)pI_s - (d_V + X)V_s$$
$$\dot{V}_r = (1 - A\epsilon_{NI})\mu pI_s + (1 - \sigma)pI_r - (d_V + X)V_r$$
$$\dot{F} = nq(E_{2,r} + E_{2,s}) + q(I_r + I_s) - d_F F$$
$$\dot{X} = rX$$
$$\dot{T} = 1$$
$$\dot{A} = 0$$
$$\dot{M} = 0$$

(a) Trajectories

Meaning	Actions
Begin NI-based treatment	*treatNI*
Begin IFN-based treatment	*treatIFN*
End the treatment	*conclude*
Inject IFN	*injectIFN*

(b) Actions

Preconditions	Action	Effects
$M = 0$	*treatNI*	$A \leftarrow 1 \wedge M \leftarrow 1$
$M = 0$	*treatIFN*	$M \leftarrow 2$
$M = 1$	*conclude*	$A \leftarrow 0 \wedge M \leftarrow 3$
$M = 2$	*conclude*	$M \leftarrow 3$
$M = 2 \wedge t_i \mid T$	*injectIFN*	$F \leftarrow F + f$

(c) Transitions

Notice that the IR model does not take into account IFN and, thus, the value estimated for the parameters that directly connected with IFN, i.e., q and ψ, may need some scaling to match the real kinetics of interferon in humans infected by H1N1. Nevertheless, our model exhibits the same IFN peek as the one proposed by Saenz *et al.* in the case of an A/equine/Kildare/89 (H3N8) infection [42].

Our hybrid model chooses one treatment between the IFN-based and the NI-based one by using either the action *treatIFN* or the action *treatNI*, respectively. In the former case, the effect of the antiviral is assumed to be constant along all the treatment; in the latter case, the host is injected with an arbitrary dose f of IFN every t_i days from the

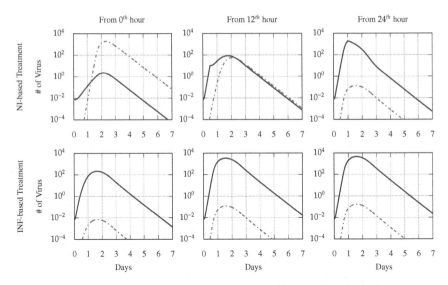

Fig. 1. A comparison between the two treatments. The plain and the dotted lines represent the number of NI-sensible virus (V_s) and the number of NI-resistant virus (V_r), respectively. Early NI-based treatment leads to a proliferation of the NI-resistant strain.

begin of the treatment until the end of it. Figure 1 represents the evolution of the two variables V_s (plain line) and V_r (dotted line) under both the considered treatments.

Notice that since the influenza model has not been obtained through composition of sub-models, it does not contain silent transitions. Hence, weak and strong bisimulations coincide, $(\xrightarrow{\tau})^* = (\xrightarrow{\tau})^0 = Id$, and $\Rightarrow^{Exp}$ means that the system is able to perform an exposed transition.

First of all, we observe that the influenza model is sensitive to treatments. As a matter of facts, it does not satisfy the weak unwinding $\mathcal{W}(Id, \approx^{Exp}, (\xrightarrow{\tau})^*)$, since for instance the system after an NI-based treatment is not bisimilar to the system before treatment.

Then, we may want to decide whether the effectiveness of two treatments depends or not on the timing. We notice that the MLTS associated to our hybrid model does not satisfy the 0.5-delayed unwinding $\mathcal{W}(\doteq^{Exp}, \preceq^{Exp}, \rightarrow^{>0.5})$. As the matter of the facts, the peek of the V_r load observable by starting the NI-based treatment at the time of the infection can be obtained neither with different timing nor by using the IFN-based treatment (see Figure 1). This proves that the effectiveness of the treatments are time-dependent.

Another question that deserves attention is whether IFN and NI are equivalent or not with respect to the virus level. We consider *treatNI* being an imposed action and verify that the MLTS associated to the proposed influenza model does not satisfies the bi-exposed unwinding $\mathcal{W}(\doteq^{Exp}, \preceq^{Exp}, \Rightarrow^{Exp})$, i.e., it is not possible to obtain the same virus load produced by the NI-treatment, whenever it has began, in the same time.

6   Conclusions

We considered biological systems represented as hybrid automata, and more in general as MLTS's, which can interact with the environment through what we call imposed actions. We introduced unwinding conditions as a formalization of the notion of robustness for such systems. Then, we proposed three instances of our general framework and exploited them in the analysis of a model of influenza kinetics. The influenza model we presented is novel and has been built merging different models proposed in the literature and fitting their parameters. Although not surprising, the results of our analysis are interesting, since they demonstrate the applicability of the proposed framework.

In [9] we compared unwinding conditions with other formalizations of robustness and we investigated on their use over both hybrid automata and the process algebra Bio-PEPA. However, in [9] we have not introduced a notion of composition. In this work composition allows us to both define the notion of environment and give a clear formal meaning to weak unwinding in terms of interactions with all the possible environments.

As future work we intend to broaden our analysis to more complex models of interactions with the environment (e.g., continuous environments) and to extend our results to other unwinding instances. In particular, the analysis of more sophisticate, possibly robust, systems would require the implementation of automatic tools for checking behavioral equivalences.

Acknowledgements. This work has been partially supported by Istituto Nazionale di Alta Matematica (INdAM).

References

1. Alur, R., Courcoubetis, C., Henzinger, T.A., Ho, P.H.: Hybrid Automata: An Algorithmic Approach to the Specification and Verification of Hybrid Systems. In: Grossman, R.L., Ravn, A.P., Rischel, H., Nerode, A. (eds.) HS 1991 and HS 1992. LNCS, vol. 736, pp. 209–229. Springer, Heidelberg (1993)
2. Alur, R., Dill, D.L.: A theory of timed automata. Theoretical Computer Science 126(2), 183–235 (1994)
3. Bartocci, E., Bortolussi, L., Nenzi, L., Sanguinetti, G.: On the robustness of temporal properties for stochastic models. In: Proc. of 2nd Int. Workshop on Hybrid Systems and Biology (HSB 2013). EPTCS, vol. 125, pp. 3–19 (2013)
4. Bossi, A., Piazza, C., Rossi, S.: Compositional information flow security for concurrent programs. Journal of Computer Security 15(3), 373–416 (2007)
5. Bossi, A., Focardi, R., Piazza, C., Rossi, S.: Bisimulation and unwinding for verifying possibilistic security properties. In: Zuck, L.D., Attie, P.C., Cortesi, A., Mukhopadhyay, S. (eds.) VMCAI 2003. LNCS, vol. 2575, pp. 223–237. Springer, Heidelberg (2002)
6. Brim, L., Vejpustek, T., Safránek, D., Fabriková, J.: Robustness analysis for value-freezing signal temporal logic. In: Proc. of 2nd Int. Workshop on Hybrid Systems and Biology (HSB 2013). EPTCS, vol. 125, pp. 20–36 (2013)
7. Casagrande, A., Piazza, C., Policriti, A., Mishra, B.: Inclusion dynamics hybrid automata. Inform. and Comput. 206(12), 1394–1424 (2008)
8. Casagrande, A., Dreossi, T., Piazza, C.: Hybrid automata and ϵ-analysis on a neural oscillator. In: Proc. of First Int. Workshop on Hybrid Systems and Biology (HSB 2012). EPTCS, vol. 92, pp. 58–72 (2012)
9. Casagrande, A., Piazza, C.: Unwinding biological systems (April 2014), http://www.dmi.units.it/%7ecasagran/resources/unwinding.pdf (submitted)

10. Chiang, K., Fages, F., Jiang, J.-H., Soliman, S.: On the hybrid composition and simulation of heterogeneous biochemical models. In: Gupta, A., Henzinger, T.A. (eds.) CMSB 2013. LNCS, vol. 8130, pp. 192–205. Springer, Heidelberg (2013)

11. Clarke, E.M., Grumberg, O., Peled, D.A.: Model Checking. MIT Press (1999)

12. Dianzani, F., Viano, I., Santiano, M., Zucca, M., Gullino, P., Baron, S.: Tissue culture models of in vivo interferon production and action. In: Human Interferon, pp. 119–131. Springer (1978)

13. Donzé, A., Ferrère, T., Maler, O.: Efficient robust monitoring for STL. In: Sharygina, N., Veith, H. (eds.) CAV 2013. LNCS, vol. 8044, pp. 264–279. Springer, Heidelberg (2013)

14. Donzé, A., Maler, O.: Robust satisfaction of temporal logic over real-valued signals. In: Chatterjee, K., Henzinger, T.A. (eds.) FORMATS 2010. LNCS, vol. 6246, pp. 92–106. Springer, Heidelberg (2010)

15. Fainekos, G.E., Pappas, G.J.: Robustness of temporal logic specifications for continuous-time signals. Theoretical Computer Science 410(42), 4262–4291 (2009)

16. Focardi, R., Gorrieri, R.: A taxonomy of security properties for process algebras. Journal of Computer Security 3(1), 5–34 (1995)

17. Focardi, R., Rossi, S.: Information flow security in dynamic contexts. In: Proc. of 15th IEEE Computer Security Foundations Workshop (CSFW 2002), pp. 307–319. IEEE Computer Society (2002)

18. Ghosh, R., Tomlin, C.J.: Lateral Inhibition through Delta-Notch Signaling: A Piecewise Affine Hybrid Model. In: Di Benedetto, M.D., Sangiovanni-Vincentelli, A.L. (eds.) HSCC 2001. LNCS, vol. 2034, pp. 232–246. Springer, Heidelberg (2001)

19. Ghosh, R., Tiwari, A., Tomlin, C.: Automated symbolic reachability analysis; with application to delta-notch signaling automata. In: Maler, O., Pnueli, A. (eds.) HSCC 2003. LNCS, vol. 2623, pp. 233–248. Springer, Heidelberg (2003)

20. Ghosh, R., Tomlin, C.: Symbolic reachable set computation of piecewise affine hybrid automata and its application to biological modelling: Delta-notch protein signalling. Systems Biology 1(1), 170–183 (2004)

21. Goguen, J.A., Meseguer, J.: Security policies and security models. In: IEEE Symposium on Security and Privacy, pp. 11–20. IEEE Computer Society Press (1982)

22. Goujon, C., Moncorgé, O., Bauby, H., Doyle, T., Ward, C.C., Schaller, T., Hué, S., Barclay, W.S., Schulz, R., Malim, M.H.: Human mx2 is an interferon-induced post-entry inhibitor of hiv-1 infection. Nature 502(7472), 559–562 (2013)

23. Handel, A., Longini, Jr. I.M., Antia, R.: Neuraminidase inhibitor resistance in influenza: assessing the danger of its generation and spread. PLoS Comput. Biol. 3(12), e240 (2007)

24. Henzinger, T.A.: The Theory of Hybrid Automata. In: Proc. of IEEE Symposium on Logic in Computer Science (LICS 1996), pp. 278–292. IEEE Computer Society Press (1996)

25. Jacobs, L.D., Cookfair, D.L., Rudick, R.A., Herndon, R.M., Richert, J.R., Salazar, A.M., Fischer, J.S., Goodkin, D.E., Granger, C.V., Simon, J.H., et al.: Intramuscular interferon beta-1a for disease progression in relapsing multiple sclerosis. Annals of Neurology 39(3), 285–294 (1996)

26. Johnson, N.P.A.S., Mueller, J.: Updating the accounts: global mortality of the 1918-1920 "Spanish" influenza pandemic. Bull. Hist. Med. 76(1), 105–115 (2002)

27. Julkunen, I., Melén, K., Nyqvist, M., Pirhonen, J., Sareneva, T., Matikainen, S.: Inflammatory responses in influenza a virus infection. Vaccine 19, S32–S37 (2000)

28. Kitano, H.: Biological robustness. Nature Reviews Genetics 5(11), 826–837 (2004)

29. Kitano, H.: Towards a theory of biological robustness. Molecular Systems Biology 3(137) (2007)

30. Lafferriere, G., Pappas, G.J., Sastry, S.: O-minimal hybrid systems. Mathematics of Control, Signals, and Systems 13, 1–21 (2000)

31. Lynch, N.A., Segala, R., Vaandrager, F.W.: Hybrid i/o automata. Inf. Comput. 185(1), 105–157 (2003)
32. Maler, O., Manna, Z., Pnueli, A.: From timed to hybrid systems. In: Huizing, C., de Bakker, J.W., Rozenberg, G., de Roever, W.-P. (eds.) REX 1991. LNCS, vol. 600, pp. 447–484. Springer, Heidelberg (1992)
33. Milner, R.: A Calculus of Communicating Systems. Springer (1980)
34. World Health Organization: Influenza fact sheet no. 211 (March 2014), http://www.who.int/mediacentre/factsheets/fs211/en/
35. Paige, R., Tarjan, R.E.: Three partition refinement algorithms. SIAM Journal on Computing 16(6), 973–989 (1987)
36. Pnueli, A.: The temporal logic of programs. In: 18th Annual Symposium on Foundations of Computer Science, Providence, Rhode Island, USA, October 31-November 2, pp. 46–57. IEEE Computer Society Press (1977)
37. Pnueli, A.: The temporal semantics of concurrent programs. In: Kahn, G. (ed.) Semantics of Concurrent Computation. LNCS, vol. 70, pp. 1–20. Springer, Heidelberg (1979)
38. Quesada, J.R., Reuben, J., Manning, J.T., Hersh, E.M., Gutterman, J.U.: Alpha interferon for induction of remission in hairy-cell leukemia. New England Journal of Medicine 310(1), 15–18 (1984)
39. Rizk, A., Batt, G., Fages, F., Soliman, S.: Continuous valuations of temporal logic specifications with applications to parameter optimization and robustness measures. TCS 412(26), 2827–2839 (2011)
40. Ryan, P.Y.A., Schneider, S.A.: Process algebra and non-interference. Journal of Computer Security 9(1/2), 75–103 (2001)
41. Sabelfeld, A., Myers, A.C.: Language-based information-flow security. IEEE Journal on Selected Areas in Communications 21(1), 5–19 (2003)
42. Saenz, R.A., Quinlivan, M., Elton, D., MacRae, S., Blunden, A.S., Mumford, J.A., Daly, J.M., Digard, P., Cullinane, A., Grenfell, B.T., et al.: Dynamics of influenza virus infection and pathology. Journal of Virology 84(8), 3974–3983 (2010)
43. Taubenberger, J.K., Morens, D.M.: 1918 Influenza: the mother of all pandemics. Emerging Infectious Diseases 12(1), 15–22 (2006), http://www.ncbi.nlm.nih.gov/pubmed/16494711, PMID: 16494711
44. Tiwari, A., Khanna, G.: Series of Abstractions for Hybrid Automata. In: Tomlin, C.J., Greenstreet, M.R. (eds.) HSCC 2002. LNCS, vol. 2289, pp. 465–478. Springer, Heidelberg (2002)
45. Walther, E., Hohlfeld, R.: Multiple sclerosis side effects of interferon beta therapy and their management. Neurology 53(8), 1622–1622 (1999)

Attractor Equivalence: An Observational Semantics for Reaction Networks[*]

Guillaume Madelaine[1,2], Cédric Lhoussaine[1,2], and Joachim Niehren[1,3]

[1] BioComputing, LIFL Lille (CNRS UMR 8022)
[2] University of Lille 1, France
[3] INRIA Lille, France

Abstract. We study observational semantics for networks of chemical reactions as used in systems biology. Reaction networks without kinetic information, as we consider, can be identified with Petri nets. We present a new observational semantics for reaction networks that we call the attractor equivalence. The main idea of the attractor equivalence is to observe reachable attractors and reachability of an attractor divergence in all possible contexts. The attractor equivalence can support powerful simplifications for reaction networks as we illustrate at the example of the *Tet-On* system. Alternative semantics based on bisimulations or traces, in contrast, do not support all needed simplifications.

1 Introduction

A reaction network is a pair that consists of a system of chemical reactions and an initial chemical solution, to which the reactions are to be applied. The kinetics of a chemical reaction defines its speed, in function of the chemical solution to which it is applied. Reaction networks can be considered as programs of a programming language, whose operational semantics describes the evolution of the initial chemical solution over time. A prominent programming language for chemical reaction networks is BioCham [5], but there are also programming language for more powerful biochemical reaction networks such as Kappa [6] and React(C) [23].

Reaction networks have three kinds of operational semantics (see e.g. [10,32]). The stochastic semantics of a reaction network is a continuous time Markov chain which for any chemical solution S and any time point t defines the probability of reaching S at t. Stochastic simulation algorithms generate the traces of a reaction network according to its stochastic semantics. The deterministic semantics of a reaction network is a system of ordinary differential equations, that tries to approximate the average concentration of the chemical solutions at any time point. It is the input of deterministic simulation algorithms. The third, qualitative semantics, is independent of any kinetics. It is given by a binary relation, which states how a chemical solution can be reduced in a nondeterministic manner to

---

[*] This work has been funded by the French National Research Agency research grant Iceberg ANR-IABI-3096.

F. Fages and C. Piazza (Eds.): FMMB 2014, LNBI 8738, pp. 82–101, 2014.

another chemical solution by applying a reaction, and thus which solutions can be reached from the initial solution.

In the following, we will stick to the qualitative semantics of reaction networks, i.e., we will ignore all kinetic informations. This is a simplification, which broadens the scope of our approach (since the kinetics are often unknown) but at the cost of lower precision. As a consequence, reaction networks can be identified with Petri nets. The same simplification was adapted in much previous work on static analysis of reaction networks (see e.g. [45,36,42]), either since the kinetics does not matter for the question under consideration, or also, since the precise kinetics are unknown.

The observational semantics of a program is usually defined on top of the operational semantics, with the objective to formalize its input-output behaviour [31,37,39]. This is done by defining a notion of program equivalence, so that equivalent programs can be exchanged by each other in any admissible context without affecting the observable behavior. The observations of the output produced by equivalent reaction networks must thus be the same in all admissible contexts (which generalize on inputs). However, there exists no notion of observational semantics for reaction networks so far. The missing ingredients are appropriate notions of observations and admissible contexts. In particular, one cannot rely on observing termination, as in all previous work on observational semantics of functional programming languages, since biological systems may change without end during an oscillation or in an equilibrium.

In this paper we propose an observational equivalence for reaction networks, that we call the *attractor equivalence*. The main idea is to *observe reachable attractors*, i.e., strongly connected components of chemical solutions so that each of them can be reached from each other. Note that whenever an attractor is reached, then the reduction must continue to loop infinitely within the attractor or terminate if the attractor is a singleton. Our semantics may also observe the *reachability of an attractor divergence*, i.e., of a chemical solution from which no attractor can be reached. A biological example of an attractor divergence is a tumor that is growing in an irreversible manner without termination. Furthermore, the observation of a chemical solution may not be able to see all informations about all molecules. What can be observed is fixed by a parameter that we call the observation function. In a biological example, one may be able to observe only fluorescent proteins under a microscope, or be unable to distinguish two different kinds of fluorescent proteins. A context of a reaction network is itself a reaction network. Rather than admitting all contexts, we fix as a parameter a subset of "admissible" contexts, by restricting the set of molecules that these contexts may affect. This corresponds for instance to biological experiments, in which only particular molecules in the environment of a cell can be added or consumed by a microfluidic device [46], but not any other molecule, in particular not those which are produced or consumed exclusively inside the cell. Finally, for a fixed observation function and a fixed set of admissible contexts, we call reaction networks *attractor equivalent*, if they have the same observations in all admissible contexts.

We also provide a set of axioms that we prove to be correct for the attractor equivalence, and justify their relevance at the example of a biological system. We have chosen a simplistic reaction network for the *Tet-On* system [21,16] with only 10 reactions. It models the addition of *Dox*-molecules to the environment of a cell (by a microfluidic device), its transport into the cell, and the expression of the fluorescent molecule GFP_a, that can be observed by a microscope. We show that we can use our axioms to reduce the *Tet-On* network into an attractor equivalent network with only 2 reactions:

$$Dox \to Dox + GFP_a \quad \text{and} \quad GFP_a \to \emptyset$$

The only molecule that the context is admitted to affect is *Dox*. In any context $nDox$ where $n \geq 1$, this reaction network system reaches a single attractor $\{nDox + mGFP_a \mid m \geq 0\}$. Note that this would not be true without the degradation reaction. The number m of GFP_a molecules can be observed in order to measure the degree of fluorescence, which may vary from 0 to infinity in the attractor. In the empty context, the above reaction network terminates immediately with the empty solution, so the singleton attractor $\{0Dox+0GFP_a\}$ is reached, of which the absence of GFP_a is observed.

Even though very simple, the application to the *Tet-On* system already illustrates the appropriateness of the attractor equivalence, in that it is sufficiently powerful to support the needed simplifications. This is in contrast to alternative notions of program equivalences for Petri net which are based either on bisimulations or on traces [26,30,19]. It turns out that 2 of our 5 axioms used for simplification are incorrect for these alternative equivalences. The problem is that these axioms change the internal cascades from the input to the output, which spoils bisimulation and alters traces. Conversely, neither the bisimilation equivalence nor the trace equivalence are included in the attractor equivalence, since attractor divergent networks may be equivalent there to non attractor divergent networks. Therefore, these alternative equivalences are indeed inappropriate.

Outline. We recall reaction networks with the qualitative semantics in Section 2 and define the attractor equivalence in Sections 3 and 4. In Section 5, we present a set of axioms of the attractor equivalence. In Section 6 we use them to simplify the detailed *Tet-On* system from [21]. Alternative equivalence notions for reaction networks are presented in Section 7 and more related work is discussed in Section 8. We then conclude with some future work in Section 9.

2   Reaction Networks

We introduce reaction networks without kinetic functions and define their qualitative operational semantics.

Let $\mathbb{N}_0$ be the set of natural numbers including 0. We fix a set *Spec* of molecular species that will be ranged over by A, B, C.

A *(chemical) solution* $s \in Sol : Spec \to \mathbb{N}_0$ is a multiset of molecules. Given natural numbers $n_1, \ldots, n_k$, we denote by $n_1 A_1 + \ldots + n_k A_k$ the solution that contains n_i molecules of species A_i for all $1 \leq i \leq k$ and 0 molecules of all other species. We will write $s \backslash A$ for the solution obtained from solution s by removing all molecules of species A, i.e., $s \backslash A(A) = 0$ and $s \backslash A(B) = s(B)$ otherwise.

A *(chemical) reaction* $r \in Sol \times Sol$ is a pair of two solutions. We write $s_1 \rightarrow s_2$ for the reaction (s_1, s_2), and say that s_1 is the solution of reactants and s_2 the solution of products. We denote by $pr_r(A) = s_2(A) - s_1(A)$ the number of molecules A produced by r. If this number is negative, then the molecules are consumed. In order to emphasize the enzymatic part of a reaction, we will write $s_1 \xrightarrow{s_3} s_2$ as a shorthand for the reaction $s_1 + s_3 \rightarrow s_2 + s_3$. It will sometimes be convenient to have a syntax for reversible reactions. To this end, we will write $s_1 \xleftrightarrow{s_3} s_2$ instead of the set with the two reactions $s_1 \xrightarrow{s_3} s_2$ and $s_2 \xrightarrow{s_3} s_1$.

Definition 1. *A reaction network is a pair $\langle s_0, R \rangle$ consisting of a solution s_0 that we call the initial solution, and a finite set of reactions R.*

Example 1. We consider the simplified *Tet-On$_{simple}$* reaction network. In this case, the set of molecular species is $Spec = \{Dox, GFP_a\}$, and the network contains the following two reactions:

$$Dox \rightarrow Dox + GFP_a \quad \text{and} \quad GFP_a \rightarrow \emptyset \tag{1}$$

The first reaction produces active green fluorescence protein GFP_a if the enzyme doxycycline (Dox) is present outside of the cell. It can also be written as $\emptyset \xrightarrow{Dox} GFP_a$. In reality, this reaction is performed by a cascade of reactions which transport Dox into the cell in order to trigger the expression of GFP_a by transcription and translation. The second reaction degrades GFP_a by consuming one GFP_a molecule at each application. Degradation is a necessary reaction to make the system work, since otherwise, it would produce unlimited amounts of GFP_a without reaching an equilibrium and thus eventually explode the cell.

The nondeterministic operational semantics of reactions is defined by the minimal binary relation $\underset{R}{\rightarrow}$ (for R being a given reaction set) that satisfies:

$$\frac{s_1 \rightarrow s_2 \in R \qquad s \in Sol}{s_1 + s \underset{R}{\rightarrow} s_2 + s}$$

Sometimes we freely write $s_1 \underset{N}{\rightarrow} s_2$ instead of $s_1 \underset{R}{\rightarrow} s_2$ if $N = \langle s, R \rangle$ for some s.

The set of solutions accessible from an initial solution s by a set of reactions R is $Acc_R(s) = \{s' \mid s \underset{R}{\rightarrow}^* s'\}$. We also sometimes write $Acc_N = Acc_R(s)$ if $N = \langle s, R \rangle$. For the simple reaction network *Tet-On* for instance, we have $Acc_{Tet\text{-}On}(Dox) = \{Dox + m\,GFP_a \mid m \in \mathbb{N}_0\}$.

Definition 2 (Attractor). *A set $S \subseteq Sol$ is an* attractor *for a set of reactions R iff $Acc_R(s) = S$ for all $s \in S$. $\mathcal{A}(R)$ denotes the set of attractors for R.*

For instance, if its initial solution contains only a molecule of Dox, the *Tet-On$_{simple}$* network has the attractor $\{Dox + m\,GFP_a \mid m \in \mathbb{N}_0\}$ since $Dox +$

$(m - 1)GFP_a \rightarrow Dox + mGFP_a$ and conversely. The same set, however, is not an attractor of $\{Dox \rightarrow Dox + GFP_a\}$, since there, one can never come back from $Dox + mGFP_a$ to $Dox + (m - 1)GFP_a$.

Note that the problem of deciding if a solution is in an attractor is equivalent to decide if the initial marking of a Petri net is a home state, which is also the problem of deciding if a Petri net is reversible, that is decidable [1].

3   Context-Less Attractor Equivalence

We define a notion of context-less attractor equivalence for reaction networks, that we will lift to a context-sensitive version in the next section.

We first define an observation function that states what can be observed from an output chemical solution. For instance, we may want to express the fact that some proteins cannot be distinguished or that some other molecules cannot be seen at all.

Definition 3. *An observation function Ω is a homomorphism from the monoid $(Sol, +, \emptyset)$ to some other monoid.*

Example 2. For the simplified Tet-On_{simple} system, we might want to observe with a microscope the number of fluorescing molecules GFP_a. In this case, we map chemical solutions to natural numbers, i.e., to the monoid $(\mathbb{N}_0, +, 0)$.

$$\Omega_{Tet\text{-}On}(mGFP_a + s) = m \qquad \text{where } GFP_a \notin s$$

Alternatively, we may also want to observe the presence or absence of GFP_a. This can be done by using the Boolean monoid $(\{0, 1\}, \max, 0)$ as follows.

$$\Omega'_{Tet\text{-}On}(mGFP_a + s) = \min(m, 1) \qquad \text{where } GFP_a \notin s$$

Or else, we might want to count both GFP_a and Dox molecules, since the latter can also be controlled experimentally (by the microfluedic device). In this case, we can use $(Sol, +, \emptyset)$ as source and target monoid as follows:

$$\Omega''_{Tet\text{-}On}(mGFP_a + nDox + s) = mGFP_a + nDox \qquad \text{where } GFP_a, Dox \notin s$$

We next define our observation predicates which will be based on the accessibility of attractors. We fix an observation function Ω and let $O \subseteq \Omega(Sol)$ be a set of observations.

Definition 4. *We say that a network $\langle s_0, R \rangle$ may converge to O if it may reach an attractor with observables O:*

$$\langle s_0, R \rangle \downarrow_O \quad \text{iff } \exists S \in \mathcal{A}(R) \text{ s.t. } S \subseteq Acc_R(s_0) \text{ and } \Omega(S) = O$$

An attractor divergence is a solution that may not converge to any set of observations. We say that a network may diverge if it can reach an attractor divergence.

$$\langle s_0, R \rangle \uparrow \quad \text{iff } \exists s \in Acc_R(s_0) \text{ s.t. } \neg \exists O. \ \langle s, R \rangle \downarrow_O$$

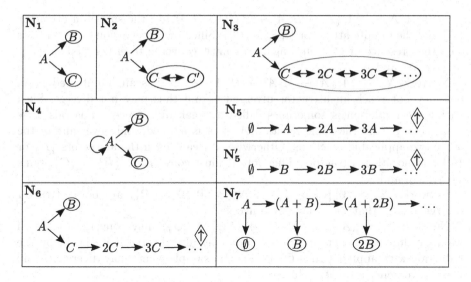

Fig. 1. Accessibility graphs of example reaction networks. The ovals denote attractors, and ↑ indicates a divergence. The networks N_1, N_2, N_3, and N_4 must converge to either $\{B\}$ or $\{C\}$ for the observation function from Example 3 which identifies C' with C and ignores the multiplicities of C's. Networks N_5 and N_5' are attractor divergences. Network N_6 may either diverge or converge to $\{B\}$. Network N_7 may converge to an infinity of observation sets $\{nB\}$ where $n \in \mathbb{N}_0$.

An attractor divergence is a solution that will change infinitely without looping. The typical example is an infinite growth of a chemical solution, which cannot be reversed by some degradation. For instance, the network $\langle \emptyset, \{\emptyset \to A\}\rangle$ is an attractor divergence, since A cannot be degraded, so that one can never come back to previous solutions but always continue, i.e., no attractor is ever reached.

Example 3. We next illustrate the observation predicates at the series of reaction networks, whose accessiblity graphs are given in Fig. 1. For these, we use the set of species $Spec = \{A, B, C, C', D\}$ and the observation function Ω that counts the number of molecules for all species, except for C and C'. For the latter, it identifies C with C' and observes only the existence of either of the two. More formally, this can be defined by $\Omega(n_1A + n_2B + n_3C + n_4C' + n_5D) = n_1A + n_2B + mC + n_5D$ where $m = \min(n_3 + n_4, 1)$.

Network $N_1 = \langle A, \{A \to B, A \to C\}\rangle$ may reach two attractors, $\{B\}$ and $\{C\}$. Hence $N_1 \downarrow_{\{B\}}$ and $N_1 \downarrow_{\{C\}}$. This network may not diverge, i.e., $\neg N_1 \uparrow$, so it must converge to either $\{B\}$ or $\{C\}$.

Network $N_2 = \langle A, \{A \to B, A \to C, C \leftrightarrow C'\}\rangle$ has two attractors $\{B\}$ and $\{C, C'\}$. Since the observation function identifies C' with C, N_2 may converge to the same sets of observables as N_1, i.e., $N_2 \downarrow_{\{B\}}$ and $N_2 \downarrow_{\{C\}}$. Furthermore, it may not diverge, so it must converge to either $\{B\}$ or $\{C\}$, same as N_1.

Network $N_3 = \langle A, \{A \rightarrow B, A \rightarrow C, C \leftrightarrow 2C\}\rangle$ has the singleton attractor $\{B\}$ and the infinite attractor $\{nC \mid n > 0\}$. Since the observation function tests only the presence of C's, once again, p_4 must converge to either $\{B\}$ or $\{C\}$, same as N_1 and N_2.

Network $N_4 = \langle A, \{A \rightarrow A, A \rightarrow B, A \rightarrow C\}\rangle$ has an infinite reduction but may not reach an attractor divergence, and thus may not diverge. Such an infinite reduction is sometimes called a "weak divergence". It is raised by unfairness, if only the same reaction $A \rightarrow A$ is applied, and never any of the other two applicable reactions. Otherwise, N_4 can reach the attractors $\{B\}$ or $\{C\}$ and no other attractors. Therefore it must converge to $\{B\}$ or $\{C\}$, same as N_1, N_2, and N_3.

Networks $N_5 = \langle \emptyset, \{\emptyset \rightarrow A\}\rangle$ et $N_5' = \langle \emptyset, \{\emptyset \rightarrow B\}\rangle$ are both attractor divergences, so that they may not converge.

Network $N_6 = \langle A, \{A \rightarrow B, A \rightarrow C, C \rightarrow 2C\}\rangle$ may converge to $\{B\}$ if we apply first the reaction $A \rightarrow B$, but may also diverge if we begin by the reduction with applying $A \rightarrow C$. This is an example where may divergence and may convergence are not exclusive.

Network $N_7 = \langle A, \{\emptyset \xrightarrow{A} B, A \rightarrow \emptyset\}\rangle$ has an infinite sequence of reductions $A \rightarrow (A+B) \rightarrow (A+2B) \ldots \rightarrow (A+nB) \ldots$, but may never diverge. This weak divergence is again raised by never applying the reaction $A \rightarrow \emptyset$ in an unfair manner. Nevertheless, N_7 may not diverge, since it may at any time point still converge to either of the $\{nB\}$ where $n \in \mathbb{N}_0$, i.e., $N_7 \downarrow_{\{nB\}}$ for all n.

The fact that our observation predicates ignore weak divergences means implicitly that we only consider fair reduction strategy, in the sense that if we can always reach some attractors, then we always reach an attractor.

Definition 5. *Given an observation Ω, we say that two reaction networks N_1 and N_2 are* context-less attractor equivalent, *denoted by $N_1 \sim_\Omega N_2$, if for all $O \subseteq \Omega(Sol)$: $N_1 \downarrow_O$ iff $N_2 \downarrow_O$ and $N_1 \uparrow$ iff $N_2 \uparrow$.*

Example 4. With one molecule of *Dox* in the initial solution, the simplified Tet-On$_{simple}$ network $\langle Dox, \{\emptyset \xrightarrow{Dox} GFP_a, GFP_a \rightarrow \emptyset\}\rangle$ is context-less equivalent to $\langle Dox, \{\emptyset \rightarrow GFP_a, GFP_a \rightarrow \emptyset\}\rangle$ with respect to Ω_{Tet-On}, since both of them must converge to $\mathbb{N}_0$. This means that they may converge to $\mathbb{N}_0$, they may not converge to any other set of observations, and they do not diverge.

Example 5. We reconsider the reaction networks from Fig. 1. The networks N_1, N_2, N_3, and N_4 are context-less equivalent with respect to the observation function chosen in Example 3 since all of them must converge to $\{B\}$ or $\{C\}$, but may not be context-less equivalent for other observation functions. Note that number of times in which the same observations can be reached does not matter.

Network N_5 and N_5' are context-less attractor equivalent even though they use different species. More generally, all attractor divergent solutions are context-less attractor equivalent.

Network N_6 is not context-less attractor equivalent to the others, since it may diverge and converge.

Network N_7 is the only one that may converge to infinitely many observation sets, so it is not context-less attractor equivalent to any other.

Let us emphasize the fact that the observation predicates $\downarrow_O$ and $\uparrow$ are sufficient to reason about other convergence and divergence predicates. For instance, a network *must converge* iff it cannot diverge. A network *must converge to O* iff it may neither diverge nor converge to any other set of observations. And a network *must diverge* iff it may not converge to any O (such as attractor divergences). So adding these three alternative observation predicates would not change our notion of context-less attractor equivalence.

4   Attractor Equivalence

We next make our attractor equivalence context dependent. The chosen class of admissible contexts should capture what kind of inputs an experimental platform can make on an biological system.

A *context* is a reaction network. The application of a context $C = \langle s, R \rangle$ to a network $N = \langle s', R' \rangle$ is defined by the union of their solutions and reactions:

$$C[N] = \langle s' + s, R' \cup R \rangle$$

However, for a given biological experiment, not all contexts are admissible. Indeed, a context may not have access to some biological species because of compartmentalization. Furthermore, some simplifications of a reaction networks may only be correct only if restricting the set of admissible contexts. There are many different manners to restrict the set of contexts to a subclass of admissible contexts. Here, we adopt the simplest method, which is to fix a set $\mathcal{I} \subseteq Spec$ of input species that can be touched by the context.

Definition 6. *A context $\langle s, R \rangle$ is admissible for $\mathcal{I}$ if all species A with $A \in s$ or $A \in R$ belong to $\mathcal{I}$.*

Definition 7 (Attractor equivalence). *We call two reaction networks N_1 and N_2 attractor equivalent and write $N_1 \equiv_{\Omega, \mathcal{I}} N_2$ iff for any context C admissible for $\mathcal{I}$ the context-less attractor equivalence $C[N_1] \sim_\Omega C[N_2]$ holds.*

Example 6. In the *Tet-On* system, the only input species that can be added by the microfluedic device is Dox, so we define $\mathcal{I}_{Tet\text{-}On} = \{Dox\}$. The simplified *Tet-On* network $N_1 = \langle Dox, \{\emptyset \xrightarrow{Dox} GFP_a, GFP_a \rightarrow \emptyset\} \rangle$ is context-less attractor equivalent to $N_2 = \langle Dox, \{\emptyset \rightarrow GFP_a, GFP_a \rightarrow \emptyset\} \rangle$ but not attractor equivalent, since $C = \langle \emptyset, \{Dox \rightarrow \emptyset\} \rangle$ is admissible, but $C[N_1]$ must converge to $\{0\}$, while $C[N_2]$ must converge to $\mathbb{N}_0$.

Example 7. We reconsider the examples in Fig. 1 with all species as input species $\mathcal{I}_1 = \{A, B, C, C', D\}$. Networks N_1 and N_2 are then not attractor equivalent, since in the admissible context $C = \langle \emptyset, \{C' \rightarrow D\} \rangle$, N_2 may converge to $\{D\}$ in contrast to N_1. The context can thus make the difference between C and C'

$$\frac{A \notin s \qquad \Omega(A) = \Omega(s) \qquad A \notin \mathcal{I}}{\langle s_0, R \cup \{s \leftrightarrow A\}\rangle \equiv_{\Omega,\mathcal{I}}^{Rev} \langle s_0[s/A], R[s/A]\rangle} \text{ (REVERSIBLE)}$$

$$\frac{s_0(A) \geq n \qquad A \notin s_1 \qquad \forall r \in R.pr_r(A) \geq 0 \qquad A \notin \mathcal{I}}{\langle s_0, R \cup \{s_1 \xrightarrow{nA} s_2\}\rangle \equiv_{\Omega,\mathcal{I}}^{Enz} \langle s_0, R \cup \{s_1 \rightarrow s_2\}\rangle} \text{ (ENZYME)}$$

$$\frac{\forall s_1 \rightarrow s_2 \in R.\ A \notin s_1 \qquad \Omega(A) = \mathbf{0} \qquad A \notin \mathcal{I}}{\langle s_0 + nA, R\rangle \equiv_{\Omega,\mathcal{I}}^{Use} \langle s_0, R\rangle} \text{ (USELESS)}$$

$$\frac{A \notin s_0 + s + s' \qquad A \notin R \qquad \Omega(A) = \mathbf{0} \qquad A \notin \mathcal{I}}{\langle s_0, R \cup \{\emptyset \xrightarrow{s} A, A \rightarrow \emptyset, \emptyset \xrightarrow{A} s'\}\rangle \equiv_{\Omega,\mathcal{I}}^{Cas1} \langle s_0, R \cup \{\emptyset \xrightarrow{s} s'\}\rangle} \text{ (CASCADE}_1)$$

$$\frac{A \notin s_0 + s + s' \qquad A \notin R \qquad \Omega(A) = \mathbf{0} \qquad A \notin \mathcal{I}}{\langle s_0, R \cup \{\emptyset \xrightarrow{s} A, A \rightarrow \emptyset, A \rightarrow s'\}\rangle \equiv_{\Omega,\mathcal{I}}^{Cas2} \langle s_0, R \cup \{\emptyset \xrightarrow{s} s'\}\rangle} \text{ (CASCADE}_2)$$

Fig. 2. Axioms of the attractor equivalence

visible, which is otherwise nonobservable. For the input species $\mathcal{I}_2 = \{A, B, D\}$, the networks N_1 and N_2 are indeed attractor equivalent.

The networks N_1 and N_4 are attractor equivalent for any set of input species, since the context cannot distinguish either how many times the same set of observations is produced by an attractor.

Contexts can be used to cure or distinguish attractor divergences. For instance, with $A \in \mathcal{I}$, and $\mathcal{C} = \langle \emptyset, \{A \rightarrow \emptyset\}\rangle$, the network $\mathcal{C}[N_5]$ can converge to $\{nA \mid n \in \mathbb{N}_0\}$. And so it is not equivalent to $N_5' = \langle \emptyset, \{\emptyset \rightarrow B\}\rangle$, since $\mathcal{C}[N_5']$ must diverge.

Example 8. If the set of input species contains "intermediate" species of a reaction cascade, then context may be used to make them visible. For instance, the networks $\langle A, \{A \rightarrow B\}\rangle$ and $\langle A, \{A \rightarrow C, C \rightarrow B\}\rangle$ are not equivalent if C is an input species. They can then be distiguished by the context $\langle \emptyset, \{C \rightarrow D\}\rangle)$

5   Axioms of Attractor Equivalence

We present 5 axioms of attractor equivalence that we will use in the next section to simplification of the *Tet-On* reaction networks from [21].

The axioms for the attractor equivalence $\equiv_{\Omega,\mathcal{I}}$ are given in Fig. 2. There we use the symbol $\mathbf{0}$ for the zero of the observations monoid to which Ω is mapping.

Axiom (REVERSIBLE) removes a reversible reaction between a molecule A and a solution s, and substitutes A with s (denoted $[s/A]$) everywhere in the network. In order to do that, A and s must have the same observations, A should not occur in s, and A should not be modifiable by the context, that is: $\Omega(A) = \Omega(s)$, $A \notin s$, and $A \notin \mathcal{I}$. For instance, this axiom can be used to simplify an enzymatic reaction network as below in Michaelis-Menten style:

$$\langle \emptyset, \{(E + S) \leftrightarrow C, C \to (E + P)\}\rangle$$

Here, an enzyme E must bind to substrate S forming a complex C, in order to produce P from S while freeing E. The assumptions for applying (REVERSIBLE) are that an admissible context cannot modify the complex C and that the complex produces the same observations than $E + S$. The above network can then be simplified to the equivalent network:

$$\langle \emptyset, \{S \xrightarrow{E} P\}\rangle$$

Axiom (ENZYME) removes an enzyme from a reaction, if it must always be present in sufficient amounts to activate the reaction, that is, if there is enough enzyme in the initial solution and if the enzyme cannot be degraded. Axiom (USELESS) removes from the initial solution all molecules of a non observable species that are never consumed by any reaction. The axioms (CASCADE$_1$) and (CASCADE$_2$) reduce an enzymatic cascade that, from a solution s, produces a solution s' mediated by a degradable molecule A. A molecule A can only be used in three reactions: a production reaction, with enzymes s ($\emptyset \xrightarrow{s} A$), a degradation reaction ($A \to \emptyset$), and a reaction ($\emptyset \xrightarrow{A} s'$ or $A \to s'$) that produces some s' provided that a molecule A is present in the solution. Moreover, A is not observable, and not modifiable by the context. Then we can replace the three previous reactions by the direct production of s' from s, that is $\emptyset \xrightarrow{s} s'$.

The following theorem states that all these axioms are sound with respect to the attractor equivalence.

Theorem 1 (Soundness). *For any $X \in \{Rev, Enz, Use, Cas_1, Cas_2\}$,*

$$\equiv^X_{\Omega, \mathcal{I}} \subseteq \equiv_{\Omega, \mathcal{I}}$$

Proof. We only show here the rather basic proof for axiom (USELESS) and the more complicated case of (CASCADE$_1$). Remember that a set S is an attractor for N iff it is closed under reduction and strongly connected, and that a network may diverge iff an attractor divergence is reachable.

Let $N \equiv^{Use}_{\Omega, \mathcal{I}} N'$ and $\mathcal{C}$ be an admissible context with respect to $\mathcal{I}$. Let $N = \langle s_0, R \cup \{s_1 \xrightarrow{nA} s_2\}\rangle$ and $N' = \langle s_0, R \cup \{s_1 \to s_2\}\rangle$. Since A cannot belong to the reactants of any reaction in R and $\mathcal{C}$, for any solution s any reaction r applicable in $s + nA$ is also applicable in s. Hence, for any solution s and s':

$$(s + nA) \xrightarrow[\mathcal{C}[N]]{}^* (s' + nA) \quad \text{iff} \quad s \xrightarrow[\mathcal{C}[N']]{}^* s'$$

Moreover, the number of A in the solutions of both networks cannot decrease. So in particular, any reachable solution in $\mathcal{C}[N]$ contains at least nA. Let S be an attractor reachable from $\mathcal{C}[N]$, it must be of the form $S = \{(s' + nA) \mid s' \in Sol\}$. Then the set $S' = \{s' \mid (s' + nA) \in S\}$ is an attractor reachable from $\mathcal{C}[N']$. Reversely, for any attractor S' reachable from $\mathcal{C}[N']$, the set $S = \{(s' + nA) \mid s' \in S'\}$ is an attractor reachable from $\mathcal{C}[N]$.

The assumption of the axiom $\Omega(A) = \mathbf{0}$ implies $\Omega(S) = \Omega(S')$. Hence for any O, $\mathcal{C}[N]$ may converge to O iff $\mathcal{C}[N']$ may converge to O.

Moreover, if a solution $s + nA$ must diverge and is reachable from $\mathcal{C}[N]$, then s must diverge and is reachable from $\mathcal{C}[N']$, and reversely. So $\mathcal{C}[N]$ may diverge iff $\mathcal{C}[N']$ may diverge, and so N and N' are attractor equivalent, that is $N \equiv_{\Omega,\mathcal{I}} N'$.

We next consider axiom (CASCADE₁). We chose an instance $N \equiv_{\Omega,\mathcal{I}}^{Cas1} N'$ of this axiom with some networks $N = \langle s_0, R \cup \{\emptyset \xrightarrow{s} A, A \to \emptyset, \emptyset \xrightarrow{A} s'\}\rangle$ and $N' = \langle s_0, R \cup \{\emptyset \xrightarrow{s} s'\}\rangle$, and a context $\mathcal{C}$ that is admissible for $\mathcal{I}$ and thus does not contain A molecules. For any solution s_1 and s_2 that do not contain A molecules, we will show that:

$$s_1 \xrightarrow[\mathcal{C}[N]]{}^* s_2 \quad \text{iff} \quad s_1 \xrightarrow[\mathcal{C}[N']]{}^* s_2$$

For the implication from left to right, we will first do the following transformations on the sequence of reduction steps justifying $s_1 \xrightarrow[\mathcal{C}[N]]{}^* s_2$. Since $A \notin s_1$, $A \notin R$ and $A \notin \mathcal{I}$, we can show that any step with $\emptyset \xrightarrow{A} s'$ must be preceded by a step with $\emptyset \xrightarrow{s} A$ but not necessarily immediately. The first transformation will move them directly after the step with $\emptyset \xrightarrow{s} A$. The resulting reduction sequence is still a valid one for $\mathcal{C}[N]$, since A cannot be used in other reactions. We will also move all steps with $A \to \emptyset$ after the previous $\emptyset \xrightarrow{A} s'$. So the transition sequence can be decomposed into the following parts: first a part without A, then the production of A by s, some productions of s' with A as an enzyme, and possibly degradations of A, then again a part without A, etc. We can then imitate the transition into $\mathcal{C}[N']$, by replacing the parts with A by $\emptyset \xrightarrow{s} s'$. For the implication from the right to the left, it is sufficient to start with a reduction sequence for $s_1 \xrightarrow[\mathcal{C}[N']]{}^* s_2$, and to replace any step $\emptyset \xrightarrow{s} s'$ by three subsequent steps with $\emptyset \xrightarrow{s} A$, $\emptyset \xrightarrow{A} s'$, and $A \to \emptyset$. We will next prove the following for all sets of observations O that:

$$\mathcal{C}[N] \downarrow_O \text{ iff } \mathcal{C}[N'] \downarrow_O .$$

Let S be an attractor reachable from $\mathcal{C}[N]$ and $s_1 \in S$. Then $s_1' = s_1 \backslash A$ is also in S, since it can be obtained by applying reaction $A \to \emptyset$ repeatedly. Let $S' = Acc_{\mathcal{C}[N']}(s_1')$. We note that S' is closed under reduction and reachable from $\mathcal{C}[N']$. In order to show that S' is strongly connected, we chose $s_2' \in S'$ and show how to reach s_1' from s_2'. Since $s_2' \in S$, we have $s_1' \xrightarrow[\mathcal{C}[N']]{}^* s_2'$. Hence $s_1' \xrightarrow[\mathcal{C}[N]]{}^* s_2'$ by the above claim (and since $A \notin s_1'$, and $A \notin s_2'$), so that s_2' and s_1' belong to attractor S for the reactions of $\mathcal{C}[N]$, so that we can conclude $s_2' \xrightarrow[\mathcal{C}[N]]{}^* s_1'$ and $s_2' \xrightarrow[\mathcal{C}[N']]{}^* s_1'$. Hence, S' is an attractor reachable from $\mathcal{C}[N']$. Since $\Omega(A) = \emptyset$, we directly have $\Omega(S) = \Omega(S')$. So if $\mathcal{C}[N] \downarrow_{\Omega(S)}$ then $\mathcal{C}[N'] \downarrow_{\Omega(S')}$ and thus $\mathcal{C}[N'] \downarrow_{\Omega(S)}$.

Conversely, let S' be a reachable attractor for $\mathcal{C}[N']$, and $s_1' \in S'$. Then s_1' is also reachable in $\mathcal{C}[N]$. Let $S = Acc_{\mathcal{C}[N]}(s_1')$. It is closed under reduction and

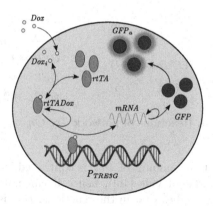

Fig. 3. Illustration of the detailed *Tet-On* network

reachable. Let $s_2 \in S$, i.e., $s_1' \xrightarrow[\mathcal{C}[N]]{}^* s_2$. Then we can apply $A \to \emptyset$ and do $s_2 \xrightarrow[\mathcal{C}[N]]{}^* s_2'$ with $s_2' = s_2 \backslash A$. So $s_1' \xrightarrow[\mathcal{C}[N]]{}^* s_2'$ implies $s_1' \xrightarrow[\mathcal{C}[N']]{}^* s_2'$, so $s_2' \in S'$, and we can do $s_2' \xrightarrow[\mathcal{C}[N']]{}^* s_1'$ and $s_2' \xrightarrow[\mathcal{C}[N]]{}^* s_1'$. So we can also do $s_2 \xrightarrow[\mathcal{C}[N]]{}^* s_1'$, so S is strongly connected and is an attractor. So if $\mathcal{C}[N'] \downarrow_{\Omega(S')}$ then $\mathcal{C}[N] \downarrow_{\Omega(S)}$ and since $\Omega(S) = \Omega(S')$, we have $\mathcal{C}[N] \downarrow_{\Omega(S')}$. It remains to show that:

$$\mathcal{C}[N] \uparrow \text{ iff } \mathcal{C}[N'] \uparrow$$

If there an attractor divergence s_1 reachable from $\mathcal{C}[N]$, then $s_1 \backslash A$ must also diverge in $\mathcal{C}[N]$, and it is reachable and must diverge in $\mathcal{C}[N']$ too. Conversely, if s_1' is an attractor divergence reachable from $\mathcal{C}[N']$, then it is directly reachable in $\mathcal{C}[N]$, and must diverge too. In summary, $\mathcal{C}[N]$ and $\mathcal{C}[N']$ are context-less attractor equivalent for all admissible contexts $\mathcal{C}$, so that $N \equiv_{\Omega, \mathcal{I}} N'$. □

6 Simplification of Detailed *Tet-On* Network

We now show how to use the axioms of the attractor equivalence for simplying the network of a concrete biological system. We have chosen the *Tet-On* system [21,15,16] that is illustrated in Fig. 3. While still being rather simple compared to other networks in biological applications, it can already be used to illustrate the power of the attractor equivalence.

The *Tet-On* system describes how the producton of activated green fluorescent proteins (*GFP_a*) in a cell can be stimulated by the presence of doxycycline (*Dox*) outside the cell. The detailed reaction network from [21] is:

$$\text{Tet-On}_{detailed} = \langle 5 P_{TRE3G} + 45000 rtTA, R^{(0)} \rangle$$

where $R^{(0)}$ is the set of reactions (2-10) from Fig. 4. The initial solution contains 5 copies of the gene P_{TRE3G} and 45000 molecules $rtTA$ (i.e. a concentration of about $1.1 \mu g/mL$).

$$\emptyset \xrightarrow{Dox} Dox_i \qquad (2)$$

$$Dox_i \to \emptyset \qquad (3)$$

$$(rtTA + Dox_i) \leftrightarrow rtTADox \qquad (4)$$

$$\emptyset \xrightarrow{P_{TRE3G}+rtTADox} mRNA \qquad (5)$$

$$mRNA \to \emptyset \qquad (6)$$

$$\emptyset \xrightarrow{mRNA} GFP \qquad (7)$$

$$GFP \to GFP_a \qquad (8)$$

$$GFP \to \emptyset \qquad (9)$$

$$GFP_a \to \emptyset \qquad (10)$$

Fig. 4. Reactions of network $Tet\text{-}On_{detailed}$

We assume here that the amount of Dox is controlled by the enviroment (for instance by a microfluidics devide [46]), hence it can not be modified by the reaction network. That implies that in the reactions, Dox is an enzyme and not a reactant. Dox can move into the cell and become Dox_i by reaction (2), where it can either be degraded by reaction (3), or bind to the artificial transcription factor $rtTA$ by reaction (4). The complex $rtTADox$ can then either dissociate (4), or activate the transcription of the gene P_{TRE3G}, producing $mRNA$ (5). $mRNA$ can either be degraded (6) or be translated into GFP (7). Finally, GFP needs to be actived into GFP_a in order to become fluorescent and thus observable by a microscope (8). Both GFP and GFP_a can also be degraded (9, 10).

We want to simplify the complex network $Tet\text{-}On_{detailed}$ to the simplified network $Tet\text{-}On_{simple}$ in (1), while consider the observation function $\Omega_{Tet\text{-}On}$ defined by $\Omega_{Tet\text{-}On}(m\,GFP_a + s) = m$ where s is an arbitrary solution with $GFP_a \notin s$ and $m \in \mathbb{N}_0$, and with the set of input species $\mathcal{I}_{Tet\text{-}On} = \{Dox\}$.

By using the axioms from Fig. 2, we will show that the networks $Tet\text{-}On_{detailed}$ and $Tet\text{-}On_{simple}$ are attractor equivalent. We start with the complex network $Tet\text{-}On_{detailed}$ and simplify it by applying the axioms. Let us first consider the reversible reaction (4). Since $\Omega_{Tet\text{-}On}(rtTA + Dox_i) = \Omega_{Tet\text{-}On}(rtTADox) = 0$ and $rtTADox \notin \mathcal{I}_{Tet\text{-}On}$, we can substitute $rtTA + Dox_i$ for $rtTADox$ by applying axiom (REVERSIBLE). This shows that $Tet\text{-}On_{detailed}$ is equivalent to the network:

$$Tet\text{-}On^{(1)} = \langle 5P_{TRE3G} + 45000rtTA, R^{(1)} \rangle$$

where $R^{(1)}$ is obtained from the $Tet\text{-}On_{detailed}$ reaction set $R^{(0)}$ by removing reaction (4) and by replacing reaction (5) by:

$$\emptyset \xrightarrow{P_{TRE3G}+rtTA+Dox_i} mRNA \qquad (5')$$

This reaction is then the only one that use $rtTA$. Since it is as an enzyme, and since $rtTA$ is present in sufficient amount in the initial solution, we can apply the axiom (ENZYME), and $rtTA$ can be removed from that reaction. Similarly, we can remove P_{TRE3G}. So the reaction becomes:

$$\emptyset \xrightarrow{Dox_i} mRNA \qquad (5'')$$

Then P_{TRE3G} and $rtTA$ become useless, and since they are unobservable, they can be removed from the initial solution using the axiom (USELESS). Network

$Tet\text{-}On^{(1)}$ is thus equivalent to network: $Tet\text{-}On^{(2)} = \langle \emptyset, R^{(2)} \rangle$ where $R^{(2)}$ contains the reactions (2),(3), (5″), and the reactions from (6-10). Now, $mRNA$ is unobservable, degradable, only producible by an enzymatic reaction (5″), and is only used as an enzyme to produce GFP (7). Thus, we can remove it using axiom (CASCADE₁). We can then apply the axiom (CASCADE₂) to remove GFP, since it is also in an enzymatic cascade. So $Tet\text{-}On^{(2)}$ is equivalent to the network $Tet\text{-}On^{(3)} = \langle \emptyset, R^{(3)} \rangle$ where $R^{(3)}$ is the set of reactions (2), (3), (8′) and (10) and:

$$\emptyset \xrightarrow{Dox_i} GFP_a \qquad (8')$$

Finally, we can again apply (CASCADE₁) to remove Dox_i which makes $Tet\text{-}On^{(3)}$ equivalent to $Tet\text{-}On_{simple}$. Since we only applied the axioms from Fig. 2 for our simplifications, Theorem 1 proves that indeed:

$$Tet\text{-}On_{detailed} \equiv_{\Omega_{Tet\text{-}On}, \mathcal{I}_{Tet\text{-}On}} Tet\text{-}On_{simple}$$

7   Bisimulation and Trace Equivalences

We now discuss alternative equivalences for reaction networks. These are based on bisimulations or traces of programs in some programming language, and can be transferred to reaction networks without particular difficulties, simply by considering reaction networks as programs.

We start with program equivalences based on bisimulations of programs, as developed for comparing concurrent programs in various calculi or programming languages [26,30,38].

In the case of reaction networks, we assume an observation function Ω and a set of input species $\mathcal{I}$. A reduction $s \rightarrow s'$ of a reaction network is silent if $\Omega(s) = \Omega(s')$. A silent action of a reaction network is a composition of finitely many silent reductions. An observable action is a composition of a silent action, a nonsilent reduction, and another silent action.

A *(weak) bisimulation* is a relation R on programs such that any two bisimilar programs have the same observation with Ω and can do the same observable actions while remaining in R. Two programs are *bisimilarity equivalent* if they are related by some bisimulation relation in any admissible context. The axioms (REVERSIBLE), (ENZYME) and (USELESS) are sound for the bisimulation equivalence. This doesn't hold for (CASCADE₁) and (CASCADE₂) though, as we will argue below. Hence, simplifications with these two axioms cannot be justified by the bisimilation equivalence, in contrast to the attractor equivalence as we showed in Theorem 1.

The argument is similar for both axioms. Consider for instance $N_1 \equiv_{\Omega,\mathcal{I}}^{Cas1} N_2$ of (CASCADE₁) with the networks following networks N_1 and N_2 whose reachability graphs are illustrated in Fig. 5:

$$N_1 = \langle C, \{C \rightarrow \emptyset, \emptyset \xrightarrow{C} B, B \rightarrow \emptyset, \emptyset \xrightarrow{B} D\} \rangle$$
$$N_2 = \langle C, \{C \rightarrow \emptyset, \emptyset \xrightarrow{C} D\} \rangle$$

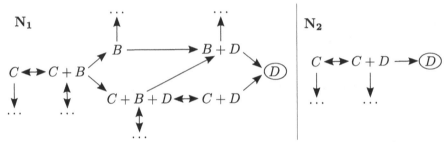

Fig. 5. N_1 *and* N_2 *are attractor equivalent (using axiom* (CASCADE$_1$)*), but nor bisimilarity equivalent@neither trace equivalent*

Fig. 6. N_3 *and* N_4 *are (weak)-bisimilar, but not attractor equivalent*

We choose the observation function $\Omega(s+nC+nD) = nC+mD$ where $C, D \notin s$ and for simplicity $\mathcal{I} = \emptyset$ so that only the empty context is admissible. The networks N_1 and N_2 are then attractor equivalent, since they both must converge to any $\{mD\}$ in all admissible contexts (since B and C can always be degraded but not D).

We next show that N_1 and N_2 are not bisimilarity equivalent. The problem is that one can reach from N_1 the nonobservable solution B, on which we can do an observable action to create observation D. In N_2, however, we need to reach C for the creation of observation D. The problem is now that B and C cannot be in any bisimilarity relation since $\Omega(B) \neq \Omega(C)$, so B cannot be in any bisimilarity relation to any solution of N_2.

Conversely, bisimilarity equivalence is not sufficient either to prove attractor equivalence. The problem is that the bisimilarity equivalence cannot distinguish some attractor divergences from some attractor convergences, when nonobservable molecules are concerned. For instance, consider the two networks in Fig. 6:

$$N_3 = \langle \emptyset, \{\emptyset \rightarrow B\} \rangle$$
$$N_4 = \langle \emptyset, \{\emptyset \rightarrow B, B \rightarrow \emptyset\} \rangle$$

Since B is nonobservable, N_3 and N_4 are in the bisimilarity equivalence. But N_3 is an attractor divergence, while N_4 may reach the attractor $\{nB \mid n \in \mathbb{N}_0\}$.

Trace equivalences [19] offer an alternative to bisimilarity equivalences. A trace is a sequence (finite or not) of solutions s_i, s.t. there is a reduction from s_i to s_{i+1} for any i. A trace observation is a sequence of observations $\Omega(s_i)$, s.t. there is an observable action from s_i to s_{i+1}. Two networks are *trace equivalent* if, in any admissible context, they have the same set of trace observations.

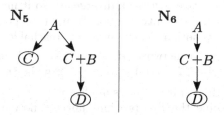

Fig. 7. N_5 and N_6 are trace equivalent, but not attractor equivalent

Once again the trace equivalence may detect differences in the intermediate behviour between input and output. In particular the axioms (CASCADE$_1$) and (CASCADE$_2$) will not preserve the trace equivalence. Note also that bisimilarity equivalence implies trace equivalence. Therefore, the axioms (REVERSIBLE), (ENZYME) and (USELESS) are also sound for the trace equivalence. Furthermore, the networks N_3 and N_4 from Fig. 6, which are not attractor equivalent, are also trace equivalent.

Beside of the problem that the trace equivalence can observe intermediate behaviors between input and output, an other problem is that it cannot detect termination or attractor convergences. But one would need to detect termination in order to distinguish the following two networks, whose reachability graph is illustrated in Fig. 7:

$$N_5 = \langle A, \{A \to C, A \to (C + B), (C + B) \to D\}\rangle$$
$$\text{and } N_6 = \langle A, \{A \to (C + B), (C + B) \to D\}\rangle$$

Networks N_5 and N_6 have the same trace observations since B is not observable, but they are not attractor equivalent, since N_5 may converge to C or D, while N_6 may only converge to D.

One could hope to solve this problem with alternative trace equivalences that are based on termination, as considered by [14,22] for instance. But since most biological system reach a cyclic attractor rather than to terminate, a terminated-trace equivalence would pose other important problems. For instance, we could slightly modify the above two networks by adding an extra loop between two fresh observable molecules, $E \leftrightarrow F$. Then the two networks would still have the same terminated-traces (none), and also the same infinite traces, but would fail to be attractor equivalent.

One could also hope to cure the notion of trace equivalence, so that it becomes included in the attractor equivalence, by restriction to some kind of fair traces. Unfortunately, it is difficult to find a good notion of fairness for Petri nets as shown in [20], were 24 different notions were proposed. Furthermore, the usual notions don't really match with what we need in systems biology or with other kinds of stochastic systems. For instance:

– if we take a fairness notion that no reaction can be ignored infinitely [20,34], then in the network $\langle A, \{A \to 2A, A \to \emptyset\}\rangle$, the trace $A \to 2A \to A \to 2A \to$

... is fair. But this trace is a weak divergence, so it may appear as unfair, since it always reach the empty solution at any time point but never does.

- if we take as fairness notion that no infinitely reachable solution is infinitely ignored [20,34], then the network $\langle A, \{\emptyset \xrightarrow{A} I, A \rightarrow B\}\rangle$ has the fair the trace $A \rightarrow (A + I) \rightarrow (A + 2I) \rightarrow (A + 3I) \rightarrow \ldots$, while the second reaction is never used (even though it is always applicable).

- if we try to combine the two previous fairness notions, i.e., we can neither ignore any reactions nor any reachable solution, then in the network $\langle A, \{A \rightarrow 2A, A \rightarrow B, \}\rangle$, that must converge, the following divergent trace is fair $A \rightarrow 2A \rightarrow (A + B) \rightarrow (2A + B) \rightarrow (A + 2B) \rightarrow (2A + 2B) \rightarrow \ldots$ (we use infinitely the 2 reactions, and the increase of B implies that there is no infinitely reachable solution)

8   Related Work

Program equivalences for sequential or concurrent programming languages are usually focused on the reachability of termination [39], which is too restrictive for reaction networks.

The notion of attractors also exists in logical regulatory networks [29,28]. These networks are quite similar to ours, but the number of each molecule is bounded, and so there are a finite number of reachable solutions, and their attractors are then finite. In [29] a simplification method for logical regulatory networks is proposed that preserves attractors, but new attractors may be created, so the reachability of the attractors may be changed. In [28], these methods are adapted so that they also preserve the reachability of attractors.

Other simplification methods have been proposed for biological systems. Plotkin [32] proposed a calculus of chemical systems, where networks are sets of reactions, without initial solution. He described then several contextual operational semantics, and some associated equivalences. In [13], the authors used subgraph epimorphisms in order to link different models. In [11], the authors used bisimulations for BioPEPA models.

Since our reaction networks can actually be seen as Petri nets, it is also interesting to look after equivalences and simplifications of Petri Nets [33]. Berthelot [4,3] has for instance proposed several simplification rules, that have been generalized for time Petri Nets [40] or for coloured Petri Nets [17]). But these rules are made to preserve usual Petri Nets properties (safety, boundedness, liveness, etc), and not the final behaviour. Other simplification methods [47,44,25] also only focus on preserving liveness and boundedness. Murata et al. also introduce new simplifications [27,24], but restricted to marked graphs, a subclass of Petri Nets. Simplifications also exist for free-choice nets, another subclass of Petri Nets [7,9]. Heiner et al. [18] also proposed analysis of biological systems represented by Petri Nets, but they are more focus on the transient behavior of the system.

Simplification methods also exist for models of biological systems, whose operational semantics accounts for time. For instance, in [8] they reduce systems of ordinary differential equations by merging some molecules, and Batmanov et al.[2] take advantage of symmetries to reduce a model while preserving stochastic properties. Other methods use difference between the speed of the reactions [35,41,12], or local equilibrium [43].

9 Conclusion

We presented the attractor equivalence for reaction networks without kinetic functions, and argued that it yields an appropriate semantics for reaction networks in systems biology. In particular we have shown that it supports powerful axioms that make it suitable for model simplification of concrete biological systems.

In future work, we plan to elaborate a more complete set of axioms, and to develop a simplification algorithm that can be applied more systematically to biological systems. It would also be nice to have a support for the simplification of reaction networks in the SBML format. We will also study the decidability of the attractor equivalence. On the scientific side, one of the problems is to add kinetics to our observational semantics. Further problems are how to develop attractor equivalences for languages of biochemical reactions such as Kappa and React(C).

References

1. Araki, T., Kasami, T.: Decidable problems on the strong connectivity of petri net reachability sets. Theoretical Computer Science (1977)
2. Batmanov, K., Kuttler, C., Lemaire, F., Lhoussaine, C., Versari, C.: Symmetry-based model reduction for approximate stochastic analysis. In: Gilbert, D., Heiner, M. (eds.) CMSB 2012. LNCS, vol. 7605, pp. 49–68. Springer, Heidelberg (2012)
3. Berthelot, G.: Checking properties of nets using transformations. In: Rozenberg, G. (ed.) APN 1985. LNCS, vol. 222, pp. 19–40. Springer, Heidelberg (1986)
4. Berthelot, G., Roucairol, G.: Reduction of petri-nets. In: Mazurkiewicz, A. (ed.) MFCS 1976. LNCS, vol. 45, pp. 202–209. Springer, Heidelberg (1976)
5. Chabrier-Rivier, N., Fages, F., Soliman, S.: The biochemical abstract machine BIOCHAM. In: Danos, V., Schachter, V. (eds.) CMSB 2004. LNCS (LNBI), vol. 3082, pp. 172–191. Springer, Heidelberg (2005)
6. Danos, V., Laneve, C.: Formal molecular biology. Theoretical Computer Science 325(1) (2004)
7. Desel, J.: Reduction and design of well-behaved concurrent systems. In: Baeten, J.C.M., Klop, J.W. (eds.) CONCUR 1990. LNCS, vol. 458, pp. 166–181. Springer, Heidelberg (1990)
8. Dokoumetzidis, A., Aarons, L.: Proper lumping in systems biology models, p. 3 (October 2009)
9. Esparza, J., Silva, M.: Top-down synthesis of live and bounded free choice nets. In: Rozenberg, G. (ed.) APN 1991. LNCS, vol. 524, pp. 118–139. Springer, Heidelberg (1991)
10. Fages, F., Soliman, S.: Formal cell biology in biocham. In: Bernardo, M., Degano, P., Zavattaro, G. (eds.) SFM 2008. LNCS, vol. 5016, pp. 54–80. Springer, Heidelberg (2008)

11. Galpin, V., Hillston, J.: Equivalence and discretisation in bio-pepa. In: Degano, P., Gorrieri, R. (eds.) CMSB 2009. LNCS, vol. 5688, pp. 189–204. Springer, Heidelberg (2009)

12. Galpin, V., Hillston, J., Ciocchetta, F.: A semi-quantitative equivalence for abstracting from fast reactions. Electronic Proceedings in Theoretical Computer Science 67(CompMod) (September 2011)

13. Gay, S., Soliman, S., Fages, F.: A graphical method for reducing and relating models in systems biology. Bioinformatics 26(18) (September 2010)

14. van Glabbeek, R.J.: The linear time-branching time spectrum. In: Baeten, J.C.M., Klop, J.W. (eds.) CONCUR 1990. LNCS, vol. 458, pp. 278–297. Springer, Heidelberg (1990)

15. Gossen, M., Bujard, H.: Tight control of gene expression in mammalian cells by tetracycline-responsive promoters. Proceedings of the National Academy of Sciences of the United States of America 89(12) (June 1992)

16. Gossen, M., Freundlieb, S., Bender, G., Müller, G., Hillen, W., Bujard, H.: Transcriptional activation by tetracyclines in mammalian cells. Science 268(5218) (June 1995)

17. Haddad, S.: A reduction theory for coloured nets. In: High-level Petri Nets, pp. 399–425 (1991)

18. Heiner, M., Gilbert, D., Donaldson, R.: Petri nets for systems and synthetic biology. In: Bernardo, M., Degano, P., Zavattaro, G. (eds.) SFM 2008. LNCS, vol. 5016, pp. 215–264. Springer, Heidelberg (2008)

19. Hoare, C.: A model for communicating sequential process (1980)

20. Howell, R., Rosier, L., Yen, H.: A taxonomy of fairness and temporal logic problems for Petri nets. Theoretical Computer Science 82, 341–372 (1991)

21. Huang, Z., Moya, C., Jayaraman, A., Hahn, J.: Using the Tet-On system to develop a procedure for extracting transcription factor activation dynamics. Molecular Bio. Systems 6(10) (October 2010)

22. Jantzen, M.: Language theory of Petri nets. In: Brauer, W., Reisig, W., Rozenberg, G. (eds.) APN 1986. LNCS, vol. 254, pp. 397–412. Springer, Heidelberg (1987)

23. John, M., Lhoussaine, C., Niehren, J., Versari, C.: Biochemical reaction rules with constraints. In: Barthe, G. (ed.) ESOP 2011. LNCS, vol. 6602, pp. 338–357. Springer, Heidelberg (2011)

24. Johnsonbaugh, R., Murata, T.: Additional methods for reduction and expansion of marked graphs. IEEE Transactions on Circuits and Systems 28(10) (1981)

25. Lee, K.-H., Favrel, J.: Hierarchical reduction method for analysis and decomposition of Petri nets. IEEE Transactions on Systems, Man, and Cybernetics SMC-15(2) (March 1985)

26. Milner, R.: A Calculus of Communication Systems. LNCS, vol. 92. Springer, Heidelberg (1980)

27. Murata, T., Koh, J.: Reduction and expansion of live and safe marked graphs. IEEE Transactions on Circuits and Systems 27(1) (1980)

28. Naldi, A., Monteiro, P.T., Chaouiya, C.: Efficient handling of large signalling-regulatory networks by focusing on their core control. In: Gilbert, D., Heiner, M. (eds.) CMSB 2012. LNCS, vol. 7605, pp. 288–306. Springer, Heidelberg (2012)

29. Naldi, A., Remy, E., Thieffry, D., Chaouiya, C.: Dynamically consistent reduction of logical regulatory graphs. Theoretical Computer Science 412(21) (May 2011)

30. Park, D.M.R.: Concurrency and automata on infinite sequences. In: Deussen, P. (ed.) GI-TCS 1981. LNCS, vol. 104, pp. 167–183. Springer, Heidelberg (1981)

31. Pitts, A.M.: Operational semantics and program equivalence. In: Barthe, G., Dybjer, P., Pinto, L., Saraiva, J. (eds.) APPSEM 2000. LNCS, vol. 2395, pp. 378–412. Springer, Heidelberg (2002)

32. Plotkin, G.D.: A calculus of chemical systems. In: Tannen, V., Wong, L., Libkin, L., Fan, W., Tan, W.-C., Fourman, M. (eds.) Buneman Festschrift 2013. LNCS, vol. 8000, pp. 445–465. Springer, Heidelberg (2013)

33. Pomello, L., Rozenberg, G., Simone, C.: A survey of equivalence notions for net based systems. In: Rozenberg, G. (ed.) APN 1992. LNCS, vol. 609, pp. 410–472. Springer, Heidelberg (1992)

34. Queille, J., Sifakis, J.: Fairness and related properties in transition systems–a temporal logic to deal with fairness. Acta Informatica 19(3), 195–220 (1983)

35. Radulescu, O., Gorban, A.N., Zinovyev, A., Lilienbaum, A.: Robust simplifications of multiscale biochemical networks. BMC Systems Biology 2 (January 2008)

36. Reddy, V.N., Mavrovouniotis, M.L., Liebman, M.N.: Petri net representations in metabolic pathways. In: Proceedings of the 1st International Conference on Intelligent Systems for Molecular Biology, pp. 328–336. AAAI (1993)

37. Sabel, D., Schauss, M.S.: Conservative concurrency in haskell. In: Proceedings of the 2012 27th Annual IEEE/ACM Symposium on Logic in Computer Science. IEEE (2012)

38. Sangiorgi, D., Kobayashi, N., Sumii, E.: Environmental bisimulations for higher-order languages. ACM Trans. Program. Lang. Syst. 33(1) (January 2011)

39. Schmidt-Schauss, M., Sabel, D., Niehren, J., Schwinghammer, J.: Observational Program Calculi and the Correctness of Translations. Rapport de recherche, Universität Frankfurt, Laboratoire d'Informatique Fondamentale de Lille - LIFL, LINKS - INRIA Lille - Nord Europe (May 2013)

40. Sloan, R.H., Buy, U.: Reduction rules for time Petri nets. Acta Informatica 33(7), 687–706 (1996)

41. Smith, N.P., Crampin, E.J.: Development of models of active ion transport for whole-cell modelling: cardiac sodium-potassium pump as a case study. Progress in Biophysics and Molecular Biology 85(2-3) (2004)

42. Soliman, S.: Invariants and other structural properties of biochemical models as a constraint satisfaction problem. Algorithms for Molecular Biology 7(1) (2012)

43. Soliman, S., Fages, F., Radulescu, O., et al.: A constraint solving approach to tropical equilibration and model reduction. In: CB-ninth Workshop on Constraint Based Methods for Bioinformatics, Colocated with CP 2013 (2013)

44. Suzuki, I., Murata, T.: A method for stepwise refinement and abstraction of Petri nets. Journal of Computer and System Sciences 27(1) (August 1983)

45. Thomas, R.: Regulatory networks seen as asynchronous automata: a logical description. Journal of Theoretical Biology 153(1), 1–23 (1991)

46. Uhlendorf, J., Miermont, A., Delaveau, T., Charvin, G., Fages, F., Bottani, S., Batt, G., Hersen, P.: Long-term model predictive control of gene expression at the population and single-cell levels. Proceedings of the National Academy of Sciences 109(35) (2012)

47. Valette, R.: Analysis of Petri nets by stepwise refinements. Journal of Computer and System Sciences 18(1), 35–46 (1979)

Petri Nets Are a Biologist's Best Friend

Nicola Bonzanni[1,2], K. Anton Feenstra[1], Wan Fokkink[1], and Jaap Heringa[1]

[1] VU University Amsterdam, De Boelelaan 1081a,
1081 HV Amsterdam, The Netherlands
{n.bonzanni,k.a.feenstra,w.j.fokkink,j.heringa}@vu.nl
[2] The Netherlands Cancer Institute, Plesmanlaan 121,
1066 CX Amsterdam, The Netherlands

Abstract Understanding how genes regulate each other and how gene expression is controlled in living cells is crucial to cure genetic diseases such as cancer and represents a fundamental step towards personalised medicine. The complexity and the high concurrency of gene regulatory networks require the use of formal techniques to analyse the dynamical properties that control cell proliferation and differentiation. However, for these techniques to be used and be useful, they must be accessible to biologists, who are currently not trained to operate with abstract formal models of concurrency. Petri nets, owing to their appealing graphical representation, have proved to be able to bridge this interdisciplinary gap and provide an accessible framework for the construction and execution of biological networks. In this paper, we propose a novel Petri net representation, tightly designed around the classic basic definition of the formalism by introducing only a small number of extensions while making the framework intuitively accessible to a biology-trained audience with no expertise in concurrency theory. Finally, we show how this Petri net framework has been successfully applied in practice to capture haematopoietic stem cell differentiation, and the value of this approach in understanding the heterogeneity of a stem cell population.

Keywords: Petri nets, biology, gene regulatory networks.

1 Introduction

Cancer causes the second largest proportion of noncommunicable (*i.e.* noninfective) disease deaths (21%) worldwide. It is projected that the annual number of deaths due to cancer will increase worldwide from 7.6 million to 13 million in 2030 [1]. Cancer and other genetic diseases are caused by abnormalities in genes which lead to changes in gene expression (*i.e.* the amount of information transcribed from genes in gene products such as RNAs and proteins) and/or in the structure of gene products. Understanding how gene expression is regulated is a necessary step to cure genetic illnesses. Gene expression, however, is a complex and tightly regulated process in which multiple regulatory elements interact concurrently to produce a global state which, in turn, determines the composite of a cell's observable characteristics (*i.e.* its phenotype, which includes the cell's shape, function, and health condition). The collection of the

F. Fages and C. Piazza (Eds.): FMMB 2014, LNBI 8738, pp. 102–116, 2014.

interactions that controls gene expression can be encoded as a network. Gene regulatory networks (GRNs) are often represented as directed graphs in which nodes represent regulatory elements and arcs represent the effect of the presence of these elements (*i.e.* expression or repression) on the target node. However, simple directed graphs are not expressive enough to encode the complex interplay between different regulatory elements. In fact, the expression (or repression) of a single gene is controlled by the coordinated effort of multiple regulatory elements which sometimes compete to achieve opposite effects. The more nodes are included in a GRN, the more it becomes prohibitive, even for the experts in the field, to mentally associate with each set of arcs the complex interplay between the source nodes that cooperate to achieve gene expression (or repression). It seems therefore convenient to encode GRNs with a formalism which represents the prerequisites of each regulatory transition in an explicit way. Petri nets with their intuitive and explicit graphical notation already serve to model biological processes [2,3,4,5] and GRN in particular. Since the basic place-transition (PT) formalism has limitations that are incompatible with fundamental GRN properties, many extensions have been proposed during the years to model GRN [6,7,8,9]. However, while these extensions enable the use of Petri nets for GRN modelling, often they also pose an insurmountable obstacle for experimental biologists that are not used to handle complex formal methods and relate better to the simplicity of the original formalism. Therefore, we decided to address the limitation of the basic PT definition using only a small number of extensions and creating a representation intuitively accessible for experimental biologists, providing a way to fold (and unfold) the new network formalism into a traditional PT network. In collaboration with a team of experimental biologists led by Dr. Berthold Göttgens at the Cambridge Institute for Medical Research, we modelled the haematopoietic stem cells differentiation using our Petri net framework. The use of this framework (described for the first time in the following Sections) was instrumental to discover the inhibitory role of Gata1 on Fli1 – two crucial genes involved in blood cell differentiation and leukemia. The biological significance of this result and its implications are discussed in depth in a companion paper [10]. This paper was recently presented at ISMB 2013 (the largest and most prestigious Bioinformatics meeting), where it received the *Award for Best Paper in Translational Bioinformatics*. We view this prize as a token of appreciation from the biological community towards a framework that aims to bridge computer science and biology with an intuitive but powerful formalism.

The current paper, rather than focusing on theoretical advancements, aims to be an experience report describing the successful application of Petri nets to real-world experimental biology. The paper is structured as follows. In Sect. 2 we define a novel Petri net representation for GRNs based on the seminal work of Chaouiya and colleagues [7]. Section 3 elaborates on how one can interpret relevant state space properties in a biological sense. In Sect. 4 we provide an experience report on the application of our framework in the experimental lab directed by Dr. Berthold Göttgens, at the Wellcome Trust Institute/MRC in

Cambridge, UK. In Sect. 5 we discuss related work, and we present our conclusions in Sect. 6.

2   Gene Regulatory Networks as Petri Nets

Gene regulatory networks are usually grounded on different assumptions relative to other biological networks such as signalling networks. Gene regulatory networks are based on two elements: a set of genes and a set of interactions between them. In turn, interactions can be either positive or negative, when a gene product has an enhancing or repressive effect, respectively, on the expression of another gene. A directed graph can not capture the cooperative interactions that are essential to correctly reproduce the behaviour observed during *in vivo* experiments. Intuitively, we want to be able to express, without ambiguity, whether a *single* gene (*de facto* its gene product) or *multiple* genes are required for a specific interaction.

An elegant way to avoid ambiguities is to use Petri nets to encode GRNs. In this framework, places represent genes, transitions represent interactions, and the marking of a place models the level of gene expression. Unfortunately, this simple construction, based on the standard definition of PT nets, conflicts with three main assumption of GRNs; in a GRN (i) the gene products (tokens) are not consumed by the interactions (transitions); (ii) interactions might have negative effects on the gene products (tokens can be removed from post-set places); (iii) the absence of a gene product (a place not marked) can be a prerequisite for an interaction. Instead of redefining enabling conditions and a marking function, we decided to build simple Petri nets modules that satisfy these assumptions by construction, and then use these modules to build larger GRNs.

Based on the above specification, and building on previous work by Chaouiya *et al.* [7], we represent each gene g in a GRN using two complementary places $\{p_g, \bar{p}_g\} \subseteq P$ where p_g represents g being expressed, and $\bar{p}_g$ represents g being repressed. The sum of tokens in p_g and $\bar{p}_g$ always equals $\mathcal{N} \in \mathbb{Z}_{>0}$, with $\mathcal{N}$ the maximum gene expression level. Each interaction is modelled by a transition. Let i be a positive interaction of the GRN. The set of genes R_i defines the gene products necessary for the occurrence of i, S_i defines the set of genes that block the occurrence of i, and g is the gene activated by the occurrence of i. Thus, it is possible to define a transition t_i modelling i such that the pre-set of t_i is

$$^\bullet t_i = \{p_r \in P \mid r \in R_i\} \cup \{\bar{p}_s \in P \mid s \in S_i\} \cup \{\bar{p}_g\}, \tag{1}$$

and the post-set of t_i is

$$t_i^\bullet = \{p_r \in P \mid r \in R_i\} \cup \{\bar{p}_s \in P \mid s \in S_i\} \cup \{p_g\}. \tag{2}$$

Intuitively, we want to enforce that t_i can be enabled if all the required gene products are available and all gene products blocking interaction i are absent. As a result of an occurrence of t_i, a token is moved from $\bar{p}_g$ to p_g, while all the tokens consumed in the places belonging to $^\bullet t_i$ are replaced by new ones.

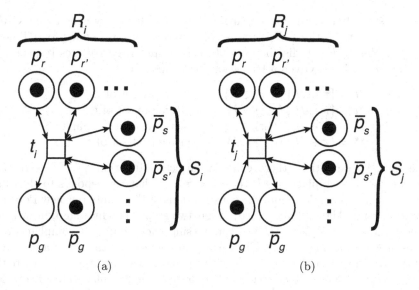

Fig. 1. Petri net modules used to model positive (a) and negative (b) gene interactions in GRNs. R is the set of genes required for the occurrence of the interaction. S is the set of genes that block the interaction. Places p represent genes in expressed state while places $\bar{p}$ represent genes in repressed state. Arcs with a double arrow head denote arcs in both directions.

This construction, depicted in Fig. 1(a), complies with assumption (i) and (iii). Similarly, we can define a transition t_j that represents a negative interaction j on a gene g by moving a token from p_g to $\bar{p}_g$. Thus, the pre-set of t_j is

$$^\bullet t_j = \{p_r \in P \mid r \in R_j\} \cup \{\bar{p}_s \in P \mid s \in S_j\} \cup \{p_g\}, \tag{3}$$

and the post-set of t_j is

$$t_j{}^\bullet = \{p_r \in P \mid r \in R_j\} \cup \{\bar{p}_s \in P \mid s \in S_j\} \cup \{\bar{p}_g\}. \tag{4}$$

This construction, depicted in Fig. 1(b), models the negative effect of j on g gene products (see assumption (ii)), and also complies with assumptions (i) and (iii). Therefore, by combining these constructions, and inferring F from the pre- and post-sets, it is possible to build a full GRN that satisfies all three assumptions using the basic PT net formalism.

One limitation of this approach is that the graphical representation of non-trivial GRNs loses its intuitiveness. Indeed, the number of places and arcs necessary to model an interesting GRN explodes. Since an intuitive graphical representation is one of our goals, we tried to compress our construction by enriching the formalism. A minimal enrichment is sufficient to achieve an elegant formalism and an intuitive graphical representation. The extended formalism includes just an additional distinction between "positive" and "negative" arcs. Hence, our new framework is defined as a tuple $B = \langle \Pi, T, F, A, I \rangle$. Π is a set of places such that there exists a single place in Π for every gene in the GRN. T is the set of transitions. F is the set of flow relations as defined for PT nets. $A \subseteq F$ and $I \subseteq F$ are disjoint sets of positive and negative arcs respectively such that $F = A \cup I$. Given a transition $t \in T$, we call

$$
\begin{aligned}
{}^\bullet t &= \{\pi \in \Pi \mid (\pi, t) \in A\} && \text{positive pre-set of } t, \\
{}^\bullet \bar{t} &= \{\pi \in \Pi \mid (\pi, t) \in I\} && \text{negative pre-set of } t, \\
t^\bullet &= \{\pi \in \Pi \mid (t, \pi) \in A\} && \text{positive post-set of } t, \text{ and} \\
\bar{t}^\bullet &= \{\pi \in \Pi \mid (t, \pi) \in I\} && \text{negative post-set of } t.
\end{aligned}
\tag{5}
$$

Note that, by definition, there exists a surjective function $\gamma : P \to \Pi$ that associates with each place $p \in P$ the place $\pi \in \Pi$ that corresponds to the same gene. Now, it is possible to fold the constructions of Fig. 1 into our new Petri net definition using Alg. 1. This algorithm has two steps. The first step, intuitively, generates the set of places Π by compressing each couple of complementary places of P into a single place. The second step generates the set of arcs A and I. For each bidirectional arc from a positive place p to a transition t we create the arc $(\gamma(p), t)$ in A. For each bidirectional arc from a negative place $\bar{p}$ to a transition t we create the arc $(\gamma(\bar{p}), t)$ in I. Finally, given the effect, positive or negative, of the transition on a gene g we create an arc $(t, \gamma(p_g))$ in A or I, respectively. The unfolding procedure is similar. Each network $B = \langle \Pi, T, F, A, I \rangle$ can be graphically represented with great parsimony of elements as shown in Fig. 2. This representation is intuitive as well as formally rigorous.

Algorithm 1 fold(N, m), where the input $N = \langle P, T, F \rangle$ is a PT Petri net marked by a vector m

---

1: $\Pi \leftarrow \emptyset$, $A \leftarrow \emptyset$, $I \leftarrow \emptyset$, m' $\leftarrow \emptyset$
2: **for all** pairs of complementary places $p_g, \bar{p}_{g'} \in P$ such that $g = g'$ **do**
3:     $\Pi \leftarrow \Pi \cup \text{new}(\pi)$ ▷ where π is a new place corresponding to gene g
4:     m'$[\pi] \leftarrow$ m$[p_g]$
5: **end for**
6: **for all** $t \in T$ **do**   ▷ see Fig. 1 notation
7:     **for all** $p_r \in P$ such that $r \in R_t$ **do**
8:         $A \leftarrow A \cup \{(\gamma(p_r), t)\}$
9:     **end for**
10:     **for all** $\bar{p}_s \in P$ such that $s \in S_t$ **do**
11:         $I \leftarrow I \cup \{(\gamma(\bar{p}_s), t)\}$
12:     **end for**
13:     **if** t models a positive interaction **then**
14:         $A \leftarrow A \cup \{(t, \gamma(p_g))\}$
15:     **else**
16:         $I \leftarrow I \cup \{(t, \gamma(p_g))\}$
17:     **end if**
18: **end for**
19: **return** $B = \langle \Pi, T, A, I, m' \rangle$

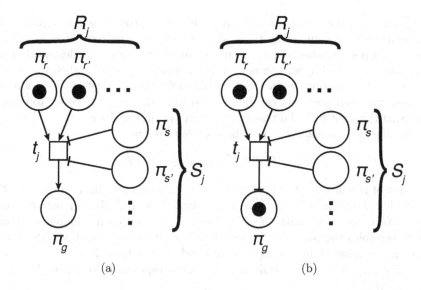

Fig. 2. Graphical representation of the positive (a) and negative (b) gene interactions depicted in Fig. 1 using the folded network definition $B = \langle \Pi, T, F, A, I \rangle$. Negative arcs belonging to I have a flat arrowhead. Both transitions are enabled by the depicted markings. In practice, R and S are small, further simplifying the representation.

Although it is possible to unfold a network in its basic PT form where the traditional enabling conditions and marking transformations apply, it is convenient to lift these conditions and transformations. Thus, given a Petri net in the form $B = \langle \Pi, T, F, A, I \rangle$, a transition $t \in T$ is said to be pre-enabled by a marking m if every place in ${}^\bullet t$ is marked by m, and every place in ${}^\bullet \bar{t}$ is *not* marked by m; it is said to be post-enabled if $\mathrm{m}(\pi) < \mathcal{N}$ for each place $\pi \in t^\bullet$, and $\mathrm{m}(\pi) > 0$ for each place $\pi \in \bar{t}^\bullet$. Finally, t is said to be enabled by a marking m if it is pre- and post-enabled by m. The occurrence of t transforms the marking m into a mapping m' defined as

$$\mathrm{m}'(\pi) = \begin{cases} \mathrm{m}(\pi) - 1 & \text{if } \pi \in \bar{t}^\bullet, \\ \mathrm{m}(\pi) + 1 & \text{if } \pi \in t^\bullet, \\ \mathrm{m}(\pi) & \text{otherwise.} \end{cases} \tag{6}$$

This formalisation can be extended to use arc weights as shown before. However, by preserving this simple definition and assigning $\mathcal{N} = 1$, it is still possible to generate biologically meaningful networks as demonstrated in [10]. These networks behave in a Boolean fashion, *i.e.* each gene can either be expressed or repressed.

In order to build biologically faithful networks, an additional fourth assumption must hold: if transcription is suspended, all gene products should, eventually, degrade over time. In Petri net terms, although tokens are not consumed by gene interactions, they must be consumed whenever the conditions that enable their production cease to hold. Intuitively, if a gene product g was produced in a previous step, but currently there are no more pre-enabled transitions that could have a positive effect on π_g, then gene product g should be degraded. This behaviour is achieved by adding new *ad hoc* transitions enabled when the conditions for gene activation are not met. Fortunately, this new set of transitions D and their pre- and post-sets can be inferred from the initial network topology, dispensing the user with the task of manually specifying them.

Formally, given a place $\pi \in \Pi$ it is possible to define

$$T_A^\pi = \{t \mid t \in T \wedge (t, \pi) \in A\} \subseteq T, \tag{7}$$

as the set of transitions that have a positive effect on π. Then, we represent T_A^π as a Boolean formula in disjunctive normal form (DNF). Each conjunctive clause of the DNF defines a transition $t \in T_A^\pi$. More specifically, each conjunctive clause uses as variables the places in $^\bullet t$, and the places in $^\bullet \bar{t}$ as negated variables. For instance, the network in Fig. 2(a) corresponds to the formula $(\pi_r \wedge \pi_{r'} \wedge \ldots \wedge \overline{\pi_s} \wedge \overline{\pi_{s'}} \wedge \ldots)$. Thus, the Boolean formula corresponding to a generic $T_A^\pi = \{t_1, t_2, \ldots, t_k\}$ is

$$\begin{aligned}
\mathcal{B}(T_A^\pi) =& (\pi_{r_{t_1}} \wedge \pi_{r'_{t_1}} \wedge \ldots \wedge \overline{\pi_{s_{t_1}}} \wedge \overline{\pi_{s'_{t_1}}} \wedge \ldots) \\
& \vee (\pi_{r_{t_2}} \wedge \pi_{r'_{t_2}} \wedge \ldots \wedge \overline{\pi_{s_{t_2}}} \wedge \overline{\pi_{s'_{t_2}}} \wedge \ldots) \\
& \vee \ldots \vee (\pi_{r_{t_k}} \wedge \pi_{r'_{t_k}} \wedge \ldots \wedge \overline{\pi_{s_{t_k}}} \wedge \overline{\pi_{s'_{t_k}}} \wedge \ldots).
\end{aligned} \tag{8}$$

Similarly, it is possible to reconstruct the network topology starting from a formula in DNF. A variable of (8) is *true* if the corresponding place is marked, *false* otherwise. Hence, if all preconditions of a transition are met, *i.e.* it is pre-enabled, then all the literals of the corresponding conjunctive clause, and the whole formula itself, evaluate as *true*. Therefore, for a generic place π, (8) evaluates as *true* if there exists a pre-enabled transition that has a positive effect on π.

$\overline{\mathcal{B}(T_A^\pi)}$, the negation of (8) in DNF, is *true* if at least one of its conjunctive clauses evaluates to *true*. By constructing D_π, the set of degradations of π, from the conjunctive clauses of $\overline{\mathcal{B}(T_A^\pi)}$, it is guaranteed that at least one of the transitions in D_π will be enabled if and only if *none* of the transitions with a positive effect on π is pre-enabled, satisfying the fourth assumption. For example, given the simple network of Fig. 3(a), $\mathcal{B}(T_A^{\pi_g})$ equals $\pi_a \wedge \pi_b \wedge \overline{\pi_c}$; therefore, $\overline{\mathcal{B}(T_A^{\pi_g})}$ is $\overline{\pi_a} \vee \overline{\pi_b} \vee \pi_c$. Fig. 3(b) shows the degradation transitions built from the conjunctive clauses of $\overline{\mathcal{B}(T_A^{\pi_g})}$.

3   State Space Analysis

For the analysis of GRNs we focus on computing the attractors of the model state space. The state space (or marking graph) is a directed multigraph where each node identifies a marking, and each arc represents the occurrence of a transition. An attractor is a forward invariant subset of the state space, *i.e.* a

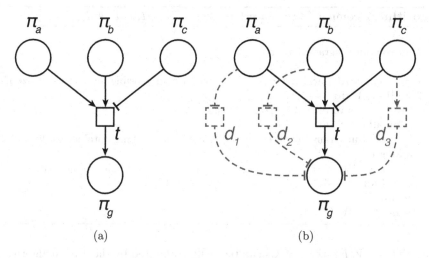

Fig. 3. The network on the right (b) shows the set of transitions D (dashed lines) necessary to model the degradation of π_g. Notice that at least one transition in D is enabled if t is not pre-enabled. Since D can be automatically inferred from the gene interactions in the initial topology (a), it is safe and usually convenient to omit the transitions of D from the graphical representation to avoid cluttering.

set such that if a marking m belongs to it, then each marking reachable from m also belongs to this set. In a biologically faithful model, each attractor should correspond to an observable biological steady state. Intuitively, as in a biological steady state the recently observed behaviour of the system will continue into the future, likewise, in the state space the model execution will keep cycling between the same states in the attractor set. Since we are particularly interested in steady states, we can abstract from time. Therefore, we can compute the state space using a fully asynchronous semantics as shown in Alg. 2. The algorithm works in two steps. First, it generates all vertices of the state space graph by computing the ensemble of all possible initial markings, *i.e.* all bit vectors of size $|\Pi|$. Second, for each bit vector the algorithm fires all enabled transitions one by one. For each transition fired, an arc from the current bit vector to the vector generated by the occurrence of the transition is added to the state space.

The strategy we choose to identify the attractors is to compute the terminal strongly connected components (TSCCs) of the state space. TSCCs are a particular class of strongly connected components (SCCs). The SCCs of a directed graph are its maximal strongly connected subgraphs, *i.e.* the induced subgraphs formed by the equivalence classes defined on the vertices by the relation of *mutual reachability*. Two vertices u and v are said to be mutually reachable if and only if there exist a path from u to v and from v to u. A TSCC T is a SCC such that if $u \in T$, then $v \in T$ for each directed arc (u, v). Thus, a TSCC is a SCC that does not have outgoing arcs to other SCCs. Therefore, by trapping the execution in a subset of states, each TSCC is an attractor of the the dynamical system. Given

---

Algorithm 2 computeStateSpace(B), where $B = \langle \Pi, T, F, A, I \rangle$

1: $E \leftarrow \emptyset$
2: $D \leftarrow$ computeDegradations(B)
3: $T \leftarrow T \cup D$
4: $V \leftarrow$ computeAllMarkings$(|\Pi|)$ ▷ compute all possible bit vectors of size $|\Pi|$
5: **for all** $v \in V$ **do**
6:     $\mathcal{E} \leftarrow$ computeAllEnabledTransitions(T, v)
7:     **for all** $e \in \mathcal{E}$ **do**
8:         $v' \leftarrow$ fireTransition(v, e) ▷ compute the marking obtained by firing e in marking v
9:             $E \leftarrow E \cup (v, v')$ ▷ add the new edge (v, v')
10:     **end for**
11: **end for**
12: **return** $G = \langle V, E \rangle$

---

a graph $G = \langle V, E \rangle$, the TSCCs can be easily computed by the Tarjan algorithm [11] in $O(|V| + |E|)$. The complexity of this analysis lies in the generation of the state space. The size of the multigraph G is exponential in the number of places; $|V| = \mathcal{N}^{|\Pi|}$, where we recall that $\mathcal{N}$ is the maximal gene expression level. For our case studies [5,10], based on the Boolean-like approach explained above, $|V|$ is 2^{11}, therefore, still tractable. Despite the complexity, building and exploring the whole state space instead of identifying single attractors may be valuable since the state space graph contains more information about the model dynamics that can be interpreted in a biological fashion. This information can lead to significative biological discoveries, as shown in [10]. However, more efficient strategies should be devised to further extend the formalism to multiple gene expression levels. Some strategies to cope with state space graphs with millions of nodes are presented in [12].

4   Modelling Haematopoietic Stem Cell Differentiation: An Experience Report

The Petri net framework defined in Sect. 2, has been specifically designed to be easily accessible to a biology-trained audience. We had the opportunity to test the applicability of our framework in the Haematopoietic Stem Cell Lab, directed by Dr. Berthold Göttgens, at the Wellcome Trust Institute/MRC in Cambridge, UK. The long term research goal of the Göttgens group is to decipher the regulatory networks responsible for blood stem cell development (*i.e.* haematopoiesis). Understanding the transcriptional control during blood cell differentiation and maturation is a necessary milestone to cure leukaemia and other types of cancer of the blood. Haematopoietic stem/progenitor cells (HSPC) have long served as a model for studying stem cells, *i.e.* cells able to differentiate, in a tree-like fashion, from a single stem (*i.e.* the stem cell) into multiple cells that accomplish different physiological functions (*i.e.* "mature" cells), such as red blood cells, white blood cells, and platelets.

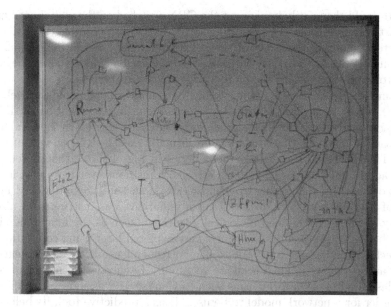

Fig. 4. The first attempt to draw the GRN that controls the haematopoietic stem cell differentiation using the Petri net formalism explained in Sect. 2. Haematopoietic Stem Cell Lab, Wellcome Trust Institute, Cambridge, UK.

The first challenge we had to face, in order to model haematopoiesis (*i.e.* blood cell differentiation), was to encode all the expertise and knowledge acquired over many years of methodical experimentation into a single formal representation. This encoding procedure is traditionally done by bioinformaticians with little experience in the application domain (in our case haematology) but good knowledge of the theoretical framework used to model biological processes. However, when the application domain is extensive and exceedingly complex, as it is often the case in biology, it becomes likely that the resulting models no longer represent faithfully the current status of the knowledge domain. Therefore, instead of trying ourselves to assimilate and interpret the enormous amount of information available in the Göttgens group or the literature, we decided to explain the basics of our framework to the experimental biologists, leveraging on the simplicity of our formalism, enabling them to encode their knowledge directly using our Petri net representation.

Although the members of the Göttgens group had no previous experience whatsoever with Petri nets or other models of concurrency, they were able to draw on the whiteboard (see Fig. 4) a first draft of the GRN that regulates the haematopoietic stem/progenitor cell differentiation in a matter of minutes. The formalisation process proved useful as well for spotting latent ambiguities in the knowledge shared within the field, such as controversial hypotheses regarding the effect of certain interactions between regulatory elements. After only a few iterative drawing and discussion cycles, the Göttgens group members were able to draw the current state of the art of the haematopoietic differentiation GRN

[10]. We could then easily implement the GRN sketched on the whiteboard using our custom-made tools. The resulting Petri net (Fig. 5) comprises 11 densely connected genes. At the heart of the network lies the triad of Scl, Gata2 and Fli1, which is characterized by extensive positive feedback loops, albeit negative regulatory interactions are common outside this central triad.

In our experience, the possibility to draw the network model on a whiteboard (Fig. 4), which then directly provides a formal and full specification of the underlying model (Fig. 5) lowered significantly the access barrier to the formalism, compared to a model that can only be formally specified using sets of equations. The consistent use of transitions nodes in Fig. 4 shows the appreciation of the additional value provided by Petri net transitions over more classic gene regulatory graphs. Being able to visualize the model on the whiteboard in the form of a graph rather than a system of equations was instrumental also to identify patterns that guided biological intuitions (such as the functional similarities between Gata2 and Fli1) which were explored later using computational approaches. Furthermore, the one-to-one correspondence between Boolean formulas and network topologies allows to rapidly re-encode the model in other formulas-based boolean formalisms such as GenYsis [13] as shown in [10].

In order for a network model to be useable as a predictive tool, its behaviour needs to be assessed using available experimental data. We therefore explored the expression patterns of the 11 regulatory genes and related these patterns to the various haematopoietic cell types. We computed the state space of our model and we performed a TSCC analysis as described in Sect. 3. Our experimentally validated network allows for three biologically sound TSCCs (*i.e.* TSCCs composed by states which markings resemble the gene expression of real blood cell types): (i) all genes are off, (ii) only Gata1 and Scl are expressed and (iii) a TSCC formed by 32 interconnected markings with multiple genes active but Gata1 always repressed. TSCC (i) corresponds to a non-haematopoietic cell. TSCC (ii) matches the erythroid cell profile. Most interesting, however, is TSCC (iii), which is composed of 32 interconnected internal states, including a state that matches the expected pattern for HSPCs. This analysis, therefore, not only demonstrates that our knowledge-driven network topology is compatible with expression patterns observed in HSPCs *in vivo*, but also suggests that the HSPC is not a homogeneous cell population; instead it is composed from cells in different stages of activation, as proved experimentally [13]. Traditional networks derived from gene expression data provide a population average which offers little insight into cellular heterogeneity and the regulatory processes likely to be critical for lineage commitment. Therefore, characterization and modelling of single cell heterogeneity is a necessary step to understand differentiation not only at intra-cellular level but also at population level, where a heterogenous population of stem cells is responsible for macroscopic changes at the level of an organism.

Analysis of transitions between different states in the model state space can be useful to predict experimental conditions for cells to differentiate out of the HSPC state. Therefore we performed a reachability analysis between each state

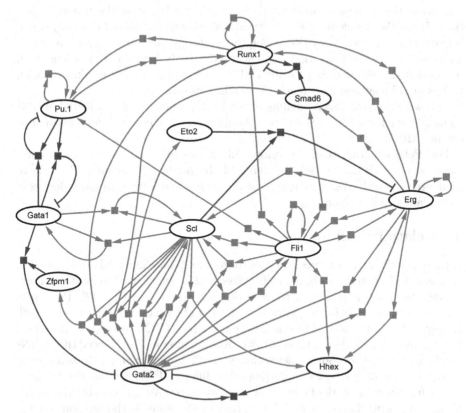

Fig. 5. Petri net of the haematopoietic gene regulatory network. Activating interactions are shown as blue arrows, repressing interactions in red with flat arrow heads. All regulatory information encoded in this model can be found in [10].

and each state matching a mature cell types in the developmental tree. Although there are no paths out of the HSPC state, which is consistent with the HSPC being a TSCC, we determined which are the closest states to the HSPC connected to a state matching a mature cell type. For example, the closest state to the HSPC that can reach the the state matching the erythrocyte cell type is at a distance of two from the HSPC, where distances are measured using the Hamming distance between state bit vectors. This observation corresponds to the notion that the transition from HSPC to erythrocyte would need at least to change the marking of two genes in the HSPC marking in order to be in a state able to reach the erythrocyte state.

Of note, experimental evidence suggests that a single change (*e.g.* ectopic expression of Gata1) would be sufficient to drive immature blood progenitors towards an erythroid fate. However, as noted above, our modelling results suggest that HSPC cells need to change at least the marking of two genes from the HSPC state. We therefore considered potentially missing network links from our current topology. In particular, we extended our model by introducing a new transition, the possible repression of Fli1 by Gata1, based on the rationale that the Fli1 regulatory element is structurally similar to the Gata2 element, which is known to be repressed by Gata1 [14]. Interestingly, just introducing this single additional repressive transition the reachability analysis of the updated network revealed that a single change in the HSPC marking (*i.e.* Gata1 marking, from 0 to 1) would be sufficient to reach the erythrocyte state.

Following on from this modelling result, an experiment was conducted in Göttgens lab which confirmed the novel inhibition on the haematopoietic progenitor cell line HPC7.

Our Petri net framework, therefore, allowed us to predict a previously unrecognized network link, which we were able to validate experimentally. Due to their preeminent biological nature, these results are discusses in more details in a companion publication [10].

5   Related Work

Formal models can be an excellent way to store and share knowledge on biological systems, and to reason about such systems [15]. Petri nets in particular are among the most used formal frameworks; for a review see [16,17]. Petri nets can be used to build both qualitative and quantitative biological models [18]. However, due to the lack of biochemical parameters, GRNs are usually described in qualitative terms. The seminal work of Kauffman *et al.* [19] showed that GRNs can be abstracted as Boolean networks, while Chaouiya *et al.* [7] showed how to encode Logical Regulatory Graphs [20] using different Petri net dialects. We built on the fundamental work of Chaouiya and colleagues. We aimed to build a simple formal representation based on PT nets that would resemble the cartoons used in experimental labs, and in which preconditions are explicitly shown (as opposed to Logical Regulatory Graphs). We followed these design principles in order to facilitate the collaboration between computer scientists and biologists and foster the use of formal methods in the experimental biology practice. Furthermore, in Sect. 2 we introduced a novel method to capture and automatically derive degradation processes in GRNs that, in our experience, proved to be crucial to faithfully simulate the regulation of gene expression.

The biological questions that drove the investigation of the haematopoietic stem cell differentiation were addressed using the state space analysis described in Sect. 3. However, it is reasonable to think that traditional analysis techniques applied to PT nets (*e.g.* T- and P-invariants analysis) could also be lifted within our framework. Some of these techniques have been shown to be useful also in the context of GRNs [8].

Although we used *ad hoc* tools for our analysis, several Petri net based tools, designed specifically for biological purposes, became recently available. Some of the most complete, popular, and stable are [21,22,23], and some interesting case studies have been already explored with these tools [8,24,25]. This is a crucial first step towards the adoption of formal models in the every day workflow of experimental labs. However, in our experience, Petri net tools can also be too general or too technical.

6   Conclusions

Biology, like computer science, is a very broad field in which each sub-community has profoundly diverse interests and a specialised terminology. Therefore, frameworks and tools that aim to be too general, targeting too many types of biological processes, tend to have an excessive complexity and therefore be unpractical. Furthermore, sometimes, tools use a computer science jargon which does not relate with the specialised vocabulary used by experimental biologists. We argue that the next generation of formalism and tools should be designed around specific biology domains (*e.g.* gene regulation, or signal transduction, or genome analysis) and their terminology should be adapted to that specific domain, allowing a natural transition from the lab bench to the computer model. To achieve this objective it will be essential to cooperate with experimental biologists through all phases of development; from the definition of the formalism to the conception of the user interface.

Acknowledgements. Parts of this work have been supported by ENFIN; a Network of Excellence funded by the European Commission within its FP6 Program, under the thematic area 'Life Sciences, genomics and biotechnology for health', contract number LSHG-CT-2005-518254.

References

1. World Health Organization: World Health Statistics 2012 (2012)
2. Koch, I., Junker, B.H., Heiner, M.: Application of petri net theory for modelling and validation of the sucrose breakdown pathway in the potato tuber. Bioinformatics 21(7), 1219–1226 (2005)
3. Gilbert, D., Heiner, M., Lehrack, S.: A unifying framework for modelling and analysing biochemical pathways using petri nets. In: Calder, M., Gilmore, S. (eds.) CMSB 2007. LNCS (LNBI), vol. 4695, pp. 200–216. Springer, Heidelberg (2007)
4. Bonzanni, N., Krepska, E., Feenstra, K.A., Fokkink, W., Kielmann, T., Bal, H., Heringa, J.: Executing multicellular differentiation: quantitative predictive modelling of c. elegans vulval development. Bioinformatics 25(16), 2049–2056 (2009)
5. Bonzanni, N., Zhang, N., Oliver, S.G., Fisher, J.: The role of proteosome-mediated proteolysis in modulating potentially harmful transcription factor activity in Saccharomyces cerevisiae. Bioinformatics 27(13), I283–I287 (2011)
6. Steggles, L.J., Banks, R., Shaw, O., Wipat, A.: Qualitatively modelling and analysing genetic regulatory networks: a petri net approach. Bioinformatics 23(3), 336–343 (2007)
7. Chaouiya, C., Remy, É., Ruet, P., Thieffry, D.: Qualitative modelling of genetic networks: From logical regulatory graphs to standard petri nets. In: Cortadella, J., Reisig, W. (eds.) ICATPN 2004. LNCS, vol. 3099, pp. 137–156. Springer, Heidelberg (2004)
8. Grunwald, S., Speer, A., Ackermann, J., Koch, I.: Petri net modelling of gene regulation of the duchenne muscular dystrophy. Biosystems 92(2), 189–205 (2008)
9. Matsuno, H., Doi, A., Nagasaki, M., Miyano, S.: Hybrid petri net representation of gene regulatory network. In: Pacific Symposium on Biocomputing, vol. 5, pp. 338–349 (2000)

10. Bonzanni, N., Garg, A., Feenstra, K.A., Schütte, J., Kinston, S., Miranda-Saavedra, D., Heringa, J., Xenarios, I., Göttgens, B.: Hard-wired heterogeneity in blood stem cells revealed using a dynamic regulatory network model. Bioinformatics 29(13), i80–i88 (2013)

11. Tarjan, R.: Depth-first search and linear graph algorithms. SIAM Journal on Computing 1(2), 146–160 (1975)

12. Krepska, E.: Towards Big Biology: High-Performance Verification of Large Concurrent Systems. PhD thesis, VU University Amsterdam (2012)

13. Garg, A., Xenarios, I., Mendoza, L., DeMicheli, G.: An efficient method for dynamic analysis of gene regulatory networks and *in silico* gene perturbation experiments. In: Speed, T., Huang, H. (eds.) RECOMB 2007. LNCS (LNBI), vol. 4453, pp. 62–76. Springer, Heidelberg (2007)

14. Grass, J.A., Boyer, M.E., Pal, S., Wu, J., Weiss, M.J., Bresnick, E.H.: Gata-1-dependent transcriptional repression of gata-2 via disruption of positive autoregulation and domain-wide chromatin remodeling. Proceedings of the National Academy of Sciences 100(15), 8811–8816 (2003)

15. Doi, A., Nagasaki, M., Matsuno, H., Miyano, S.: Simulation-based validation of the p53 transcriptional activity with hybrid functional petri net. Silico Biology 6(1), 1–13 (2006)

16. Will, J., Heiner, M.: Petri nets in biology, chemistry, and medicine - bibliography. Technical Report 04/2002, BTU Cottbus, Computer Science (2002)

17. Chaouiya, C.: Petri net modelling of biological networks. Briefings in Bioinformatics 8(4), 210 (2007)

18. Heiner, M., Gilbert, D., Donaldson, R.: Petri nets for systems and synthetic biology. In: Bernardo, M., Degano, P., Zavattaro, G. (eds.) SFM 2008. LNCS, vol. 5016, pp. 215–264. Springer, Heidelberg (2008)

19. Kauffman, S., Peterson, C., Samuelsson, B., Troein, C.: Random boolean network models and the yeast transcriptional network. Proceedings of the National Academy of Sciences of the United States of America 100(25), 14796–14799 (2003)

20. Thomas, R.: Regulatory networks seen as asynchronous automata: A logical description. Journal of Theoretical Biology 153(1), 1–23 (1991)

21. Heiner, M., Herajy, M., Liu, F., Rohr, C., Schwarick, M.: Snoopy – A unifying petri net tool. In: Haddad, S., Pomello, L. (eds.) PETRI NETS 2012. LNCS, vol. 7347, pp. 398–407. Springer, Heidelberg (2012)

22. Gonzalez, A.G., Naldi, A., Sánchez, L., Thieffry, D., Chaouiya, C.: Ginsim: A software suite for the qualitative modelling, simulation and analysis of regulatory networks. Biosystems 84(2), 91–100 (2006)

23. Nagasaki, M., Saito, A., Jeong, E., Li, C., Kojima, K., Ikeda, E., Miyano, S.: Cell illustrator 4.0: A computational platform for systems biology. Silico Biology 10(1-2), 5–26 (2010)

24. Marwan, W., Rohr, C., Heiner, M.: Petri nets in snoopy: A unifying framework for the graphical display, computational modelling, and simulation of bacterial regulatory networks. In: Helden, J., Toussaint, A., Thieffry, D. (eds.) Bacterial Molecular Networks. Methods in Molecular Biology, vol. 804, pp. 409–437. Springer (2012)

25. Doi, A., Nagasaki, M., Matsuno, H., Miyano, S.: Simulation-based validation of the p53 transcriptional activity with hybrid functional petri net. Silico Biology 6(1), 1–13 (2006)

50 Shades of Rule Composition

From Chemical Reactions to Higher Levels of Abstraction

Jakob Lykke Andersen[1], Christoph Flamm[2],
Daniel Merkle[1], and Peter F. Stadler[2−7]

[1] Department of Mathematics and Computer Science
University of Southern Denmark, Denmark
{daniel,jlandersen}@imada.sdu.dk
[2] Institute for Theoretical Chemistry, University of Vienna, Austria
xtof@tbi.univie.ac.at
[3] Bioinformatics Group, Department of Computer Science,
and Interdisciplinary Center for Bioinformatics, Leipzig, Germany
[4] Max Planck Institute for Mathematics in the Sciences, Leipzig, Germany
[5] Fraunhofer Institute for Cell Therapy and Immunology, Leipzig, Germany
[6] Center for Non-coding RNA in Technology and Health
University of Copenhagen, Denmark
[7] Santa Fe Institute, USA
studla@bioinf.uni-leipzig.de

Abstract. Graph rewriting has been applied quite successfully to model chemical and biological systems at different levels of abstraction. A particularly powerful feature of rule-based models that are rigorously grounded in category theory, is, that they admit a well-defined notion of rule composition, hence, provide their users with an intrinsic mechanism for compressing trajectories and coarse grained representations of dynamical aspects. The same formal framework, however, also allows the detailed analysis of transitions in which the final and initial states are known, but the detailed stepwise mechanism remains hidden. To demonstrate the general principle we consider here how rule composition is used to determine accurate atom maps for complex enzyme reactions. This problem not only exemplifies the paradigm but is also of considerable practical importance for many down-stream analyses of metabolic networks and it is a necessary prerequisite for predicting atom traces for the analysis of isotope labelling experiments.

1 Introduction

Abstract rule based systems with roots in process algebras and concurrency have been introduced to formalize biological processes more than a decade ago starting with Fontana's models of evolving constructive in λ-calculus [1] and Regev's view on "cells as computation" [2]. Kappa [3], for instance, was designed to model the behaviour of mixtures of "agents" (usually thought of as interacting proteins) using rules that describe changes to the agents' internal states. Conceptually similar approaches have been taken in BioNetGen [4]. Compartmentalization

F. Fages and C. Piazza (Eds.): FMMB 2014, LNBI 8738, pp. 117–135, 2014.
© Springer International Publishing Switzerland 2014

and the relative placement of cellular components is the focus of "membrane computation" [5] or the brane calculus [6]. A term-rewriting-like model has been proposed to account for the computational aspects of epigenetics [7]. For reviews see also [8,9].

Chemistry has motivated the "Chemical Abstract Machine" [10] as model of concurrent computation already in 1990, and graph representations of molecules have been used throughout the entire history of organic chemistry. Concrete models of chemistry in terms of graph rewrite systems, however, have appeared only after the turn of the millenium [11] and a systematic investigation into the practical advantages provided by the rich formal structure of graph rewriting systems is even younger. In this context it is interesting to note that Kappa has an intuitive graphical interpretation as single pushout (SPO) rewriting system on a category of suitably annotated graphs known as Σ-graphs [12]. In the context of chemistry, the more restrictive double pushout (DPO) framework appears to have some advantages. It ensures, e.g., the reversibility of chemical reactions.

A key tool in rule-based calculi is the concept of rule composition. It allows, in particular, different levels of coarse graining in the description of a system's trajectories by contracting transitions between states in a principled manner, explored in some detail for Kappa in [12]. A natural application of the same idea in the realm of chemistry is to relate elementary reactions with overall reactions composed of multiple sequential steps. Here we show rule composition is a useful avenue into disentangling the mechanistic details of multi-step transitions between a known initial and finite state. A practical problem from computational chemistry will serve as an application showcase in the following. Similar analyses could likely be useful, e.g., in generative models of animal development [13], however neither data nor rule sets are available in the form of a publicly accessible database.

The atom map of a chemical reaction specifies which atom of the product molecules corresponds to which of the educt atoms. The atom maps are a crucial piece of knowledge for the inference of metabolic fluxes from isotope-label experiments [14,15] for systems-oriented research in microbes. If lumped metabolic maps are used the experimental data alone is not sufficient to uniquely elucidate the metabolic flux pattern. In particular if model reduction is applied to genome-scale metabolic network reconstructions [16,17] automated methods to handle the atom maps properly are indispensable. This type of detailed information is in most cases not available in chemical reaction databases, so that it must be reconstructed computationally from the known educts and products only. In general, there is more than one plausible atom map and elaborate experimental techniques marking individual atoms in the educts by rare isotopes are necessary to distinguish between different possibilities. The MACiE database in addition provides information on the reaction mechanisms that are catalysed by the enzymes involved in the overall reaction [18,19]. We show here how these rules together with the knowledge of start and end state can be used to (nearly) uniquely determine the step-wise reaction mechanism. To this end we employ the formal framework of rule composition. Conversely, the alternative atom maps identify variant reaction mechanism that can be disentangled by suitable isotope labelling experiments.

The paper is structured as follows. In Section 2 we will introduce the Double Pushout formalism. Graph grammar rule composition operators will be defined and exemplified by composition of chemical reactions. In Section 3 rule composition will be employed in order to change the level of abstraction for three different chemical systems. Furthermore, full atom traces will be computed. Conclusions are given in Section 4.

2   Methods

2.1   Rule Composition

We model molecules as labelled simple graphs. Vertex labels denote atom types and edge labels indicate bond types. Chemical reactions are described by graph transformation rules in the Double Pushout (DPO) formalism that encode specific reaction mechanisms [11]. Each rule has the form $(L \xleftarrow{l} K \xrightarrow{r} R)$, where L and R are the left and right graph and K is the context graph glueing the transformation of L into R. A rule is applied to a graph G by finding a subgraph of G isomorphic to L and using the morphisms l and r to remove $L \backslash K$ from G and adding $R \backslash K$ to it, resulting in the transformed graph H. A chemical transformation rule has the special property that no atom (vertex) can be added, removed, or relabelled so that all atoms are represented in K and the restrictions r_V and l_V of the morphisms r and l to the vertex sets are bijective. Thus $\alpha = l_V \circ r_V^{-1}$ is a bijection between the vertex sets of L and R, which encodes the atom-map of the reaction mechanism. The full atom-map is obtained by extending α by the identity on the parts of G that remain unchanged in H. For a more technical description we refer to [20].

In [20] we also described in detail how transformation rules can be composed in a chemically relevant manner. Here we introduce new composition operators and give a conceptual and less formal overview of the different types of compositions and their usage. We use $\circ$ to denote composition. In contrast to the usual notation we read compositions from left to right, i.e., $r_1 \circ r_2$ is the composition which applies r_1 first and then r_2. Two rules $r_1 = (L_1 \leftarrow K_1 \rightarrow R_1)$, $r_2 = (L_2 \leftarrow K_2 \rightarrow R_2)$ are composed as $r_1 \circ_m r_2$ using a partial isomorphism m between R_1 and L_2 (note that not all partial isomorphisms m induce a valid composition). In the following we will describe types of composition for different levels of generality classified by the structure of the match m. Let $r_1 \circ_{\supseteq} r_2$ denote a composition where the match specifies that R_1 is a super-graph of L_2. Fig. 1 illustrates this case, which is analogous to the application of r_2 to the graph R_1.

Molecules correspond to connected components of molecule graphs, and connected components of the graphs in a transformation rule thus correspond to the possibility of merging and splitting molecules. The composition with a super-graph isomorphism can be generalised to a partial component-wise super-graph relation as described in [20]. In such a composition $r = r_1 \circ_{\supseteq}^c r_2$, illustrated

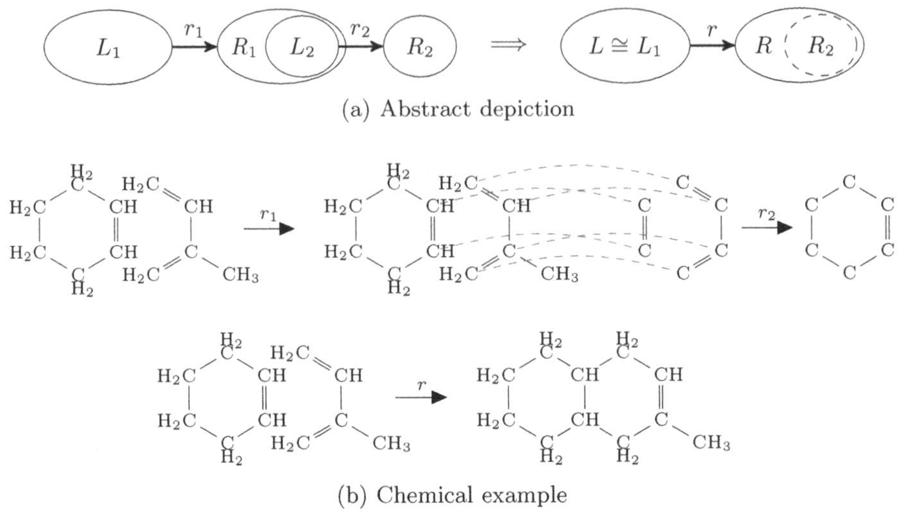

(a) Abstract depiction

(b) Chemical example

Fig. 1. A composition $r = r_1 \circ_{\supseteq} r_2$ with the matching morphism being a super-graph isomorphism of R_1 to L_2. The context graphs are omitted from the drawings for simplicity. (a) Abstract depiction; L_2 is isomorphic to a subgraph of R_2. (b) Chemical example; $r_1 = (G, G, G)$ is the identity rule for a graph G encoding the educts cyclohexene and isoprene. The second rule, r_2, is the reaction template for the Diels-Alder reaction. The composed rule therefore encodes the overall rule of the Diels-Alder reaction on the input molecules.

in Fig. 2, we only require a subset of the connected components of L_2 to be defined by the matching morphism. These components, however, still must be completely defined. The semantics of this class of composition is analogous to partial function application in programming languages.

At the most general level we consider compositions without restriction on the match. We denote these by $\circ_{\cap}$ as the match specifies a common subgraph of R_1 and L_2 (see Fig. 3 for an example). The class of chemically valid transformation rules is not closed under this type of composition. For instance, valence restrictions may be violated. As a special case we consider the composition where the common subgraph of R_1 and L_2 is empty (denoted by $\circ_{\emptyset}$). The resulting composed rule encodes the parallel application of the operand rules, as illustrated in Fig. 4.

In the following sections we will primarily use compositions with the super-graph relation, $\circ_{\supseteq}$, and we will therefore simply use $\circ$ to denote these compositions, which we refer to as "full composition". The more relaxed composition, $\circ_{\supseteq}^c$, will be referred to as "partial composition". Implicitly, we use parallel composition, $\circ_{\emptyset}$, for constructing identity rules for multiple molecules, and it will be explicitly used for the carbon-tracing in the glycolysis pathway. It should be noted that a core sub-procedure of composition with the super-graph relation is enumeration of subgraph isomorphisms, which in the general case already is NP-hard when a single morphism is desired [21]. Chemically valid transformation rules have special structure (e.g., they have bounded degree), and thus a

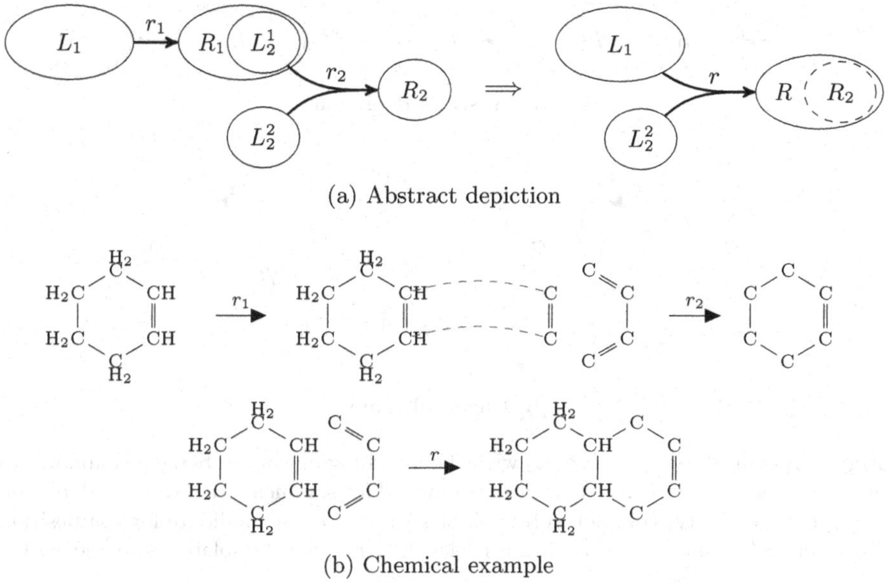

(a) Abstract depiction

(b) Chemical example

Fig. 2. A composition $r = r_1 \circ_{\supseteq}^{c} r_2$ with the matching morphism being a super-graph isomorphism of R_1 to a non-empty subset of the connected components of L_2. The context graphs are omitted for simplicity. (a) Abstract depiction; connected components of L_2 are either completely unmatched or completely matched. (b) Chemical example; r_1 is the identity rule cyclohexene and r_2 the the Diels-Alder reaction template. The composed rule encodes the partial application of the Diels-Alder reaction to the molecule, leaving the diene to be instantiated at a later stage.

better theoretical run-time may potentially be proved. This detailed analysis is however out of scope for this study, and we will pragmatically observe that the run-time in practice is not a problem for our use.

2.2 Implementation

The rule composition framework is implemented in C++11 as a part of a library that is primarily aimed at chemical graph transformation and thus includes special features and optimizations for molecules (e.g., use of canonical SMILES strings for graph isomorphism [22,23]). It is, however, not restricted to the domain of chemistry and can be used readily to handle graph grammar models at other levels of abstraction. Python bindings for the library gives a more accessible interface which allows for intuitive usage. The rule composition operators are syntactically implemented in a manner very similar to the mathematical presentation outlined in this contribution. This is achieved by using operator overloading. The computational runtime for all the experiments presented is below five minutes.

MACiE (Mechanism, Annotation, and Classification in Enzymes) [19] is a publicly available hand-curated database of enzymatic reaction mechanisms,

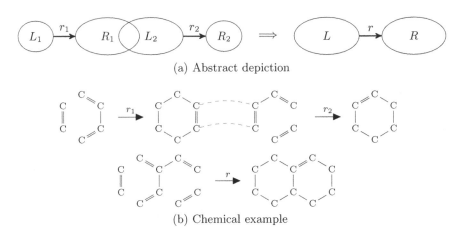

(a) Abstract depiction

(b) Chemical example

Fig. 3. A composition $r = r_1 \circ_\cap r_2$ with the matching morphism being a common subgraph R_1 and L_2. The context graphs are omitted for simplicity. (a) Abstract depiction; any (possibly empty) common subgraph of R_1 and L_2 is a candidate for composition. (b) Chemical example with both the Diels-Alder reaction template composed with it self.

Fig. 4. Composition $r = r_1 \circ_\emptyset r_2$ with an empty match, giving a composed rule which does the operand transformations in parallel

where the individual steps of the overall enzyme reaction have been experimentally verified. Detailed stepwise mechanistic information (in pictorial form) for more than 300 overall enzyme reactions can be accessed. However, atom traces for the overall enzyme reactions are not available, and information of the mechanism's flexibility with respect to a reordering of individual steps to achieve a given overall reaction is not included. Reactions in MACiE are a natural candidates for our DPO-based rule composition framework.

In the following we consider three rather complex chemical transformations as showcase examples, one example coming from the MACiE database. Each case takes up residence at a different level of organizational abstraction. First, an overall enzyme reaction mechanism is constructed from elementary reaction steps. Second, glycolysis as an example for a complete biochemical pathway with multiple split and merge points that is lumped into a single overall reaction. Finally, the formose process, an auto-catalytic reaction mechanism, illustrates that our methodology works also in networks containing cycles. We emphasize that the full atom traces for each case are computed automatically without additional external information.

3   Results

3.1 β-Lactamase

β-lacatamases (MACIE entry 0002, EC number 3.5.2.6) are bacterial enzymes that convey resistance against β-lactame antibiotics such as penicillins by catalysing the overall reaction

$$\underset{\text{(CHEBI:35627)}}{\beta\text{-Lactam}} \quad + \quad \underset{\text{(CHEBI:15377)}}{\text{water}} \quad \rightarrow \quad \underset{\text{(CHEBI:33705)}}{\text{substituted } \beta\text{-amino acid}}$$

by means of a 5-step mechanism, which is detailed in MACiE as follows (see database entry for full details): (1) Lys73 deprotonates Ser70 thereby initiating a nucleophilic addition onto the carbonyl carbon of the β-lactam. (2) The resulting intermediate collapses, cleaving the C-N bond of the β-lactam and the nitrogen deprotonates Ser130. (3) Ser130 deprotonates Lys73. (4) Glu166 deprotonates water, which initiates a nucleophilic addition at the carbonyl carbon. (5) Collapse of this intermediate leads to cleavage of the acyl-enzyme bond and liberates Ser70, which in turn deprotonates the Glu166. The 5 individual steps were modelled as graph grammar rules $r_1, \ldots, r_5$ depicted in Fig. 5. For step (2) an alternative mechanism has been suggested in [24]: protonation of the β-lactam nitrogen occurs as the first step in the reaction as an initiation step and not as a consequence of the C-N bond cleavage. We modelled this alternative as a replacement of rule r_2 by two graph grammar rules r_{1b} and r_{2b}, see Fig. 6.

The atom traces for the overall reaction is computed by a composition of the rules $r_1, \ldots, r_5$ with the identity rule for the input compounds, i.e., the β-lactam, water, and the catalysts (Glu166, Lys73, and twice Ser130). Let G and H be the the graph representation of the input and output compounds, respectively. By $\imath_G = (G, G, G)$ and $\imath_H = (H, H, H)$ we denote the corresponding identity rules. The overall composition

$$\imath_G \circ r_1 \circ r_2 \circ r_3 \circ r_4 \circ r_5 \circ \imath_H \tag{1}$$

results in the two overall rules depicted in Fig. 7. Both are in agreement with the overall mechanism given in MACiE and differ only in their hydrogen traces. The overall cyclic virtual transition states are an 8 cycle and a 10 cycle, which only differ by the exchange of a hydrogen in the amino group of Glu. The alternative model for step 2, which corresponds to

$$\imath_G \circ r_1 \circ r_{1b} \circ \circ r_{2b} \circ r_3 \circ r_4 \circ r_5 \circ \imath_H \tag{2}$$

results in the same two overall rules.

In order to check the flexibility of the reaction with respect to the order of the individual steps of the enzyme mechanism, we investigated all permutations of the rules for the composition order and verified whether the resulting overall rule produces the substituted β-amino acid as final product. Formally, we compute

$$\imath_G \circ r_{\sigma(1)} \circ \ldots r_{\sigma(5)} \circ \imath_H$$

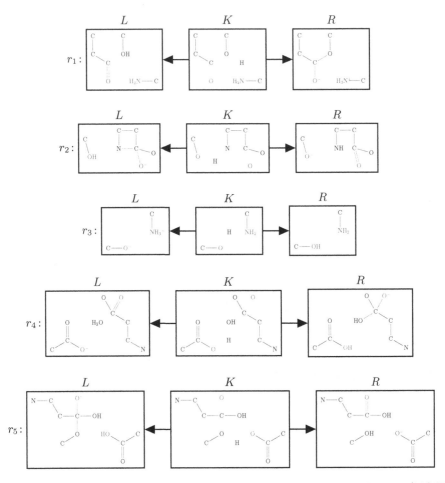

Fig. 5. β-lacatamase: transformation rules for the 5-step enzyme mechanism (MACIE entry 0002, EC number 3.5.2.6)

for all 120 permutations σ. Only the following three compositions are well-defined and result in the expected overall rules: $(r_1, r_2, r_3, r_4, r_5)$, $(r_1, r_2, r_4, r_3, r_5)$, and $(r_1, r_2, r_4, r_5, r_3)$. A detailed inspection shows that step r_3 is the recycling step of the mechanism, which can be applied concurrently to steps r_4 and r_5.

The same experiment based on the rule set $\{r_1, r_{1b}, r_{2b}, r_3, r_4, r_5\}$ shows that eight compositions are possible, all resulting in the same atom traces as given above. The first two steps need to be r_1 and r_{1b}, their relative order however is arbitrary. The subsequent rules r_{2b}, r_4, and r_5 must be in this order. The recycling step r_3 requires the rules r_1 and r_{1b} as prerequisite, but can be performed concurrently to the remaining steps, i.e., it may appear in position 3, 4, 5, or 6, thus accounting for the 8 feasible permutations.

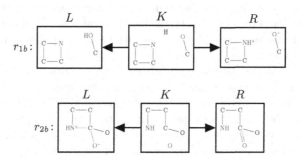

Fig. 6. β-lacatamase: transformation rules to replace step r_2 from Fig. 5, based on the mechanism as suggested in [24]

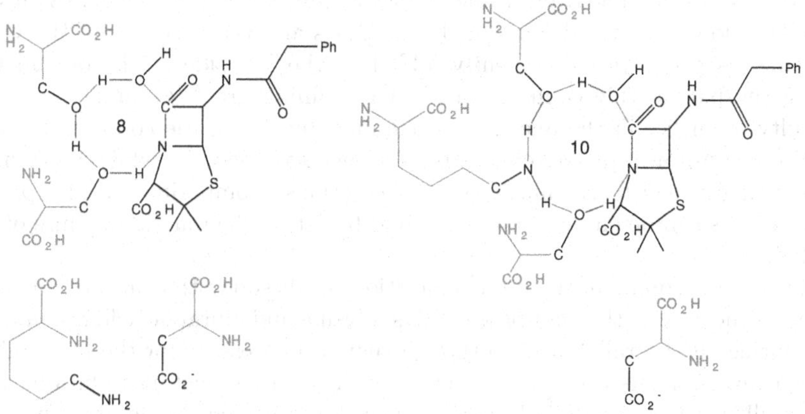

Fig. 7. The two overall reactions resulting from composing the graph rewrite rules for the elementary steps of the β-lactamase EC 3.5.2.6 (MACiE entry 0002), cmp. Eq. (1) (left, 5 steps) and Eq. (2) (right, 6 steps). Red bonds are broken while green bonds are formed during the transformation. While the overall reactions (as typically found in metabolic databases such as KEGG or MetaCyc) are identical, they differ in their hydrogen trace and the size (8 or 10) of the cyclic virtual transition state. Note that the acid/basic catalysts (the two amino acids lysine and glutamine) needed for the reaction to work still show up as precondition in the overall rules. Using partial composition results in two more generic overall reactions. These two rules are depicted as the strict subgraphs resulting from removing the gray parts from the catalysts.

This method allows for an automated analysis of the flexibility of the ordering of individual steps. Note, that usually a relatively small number of all possible permutations have to be computed, as most often already the composition of a prefix of an arbitrarily chosen permutation is not possible. For instance, in the previous example, only two of the 30 possible initial two steps are feasible, which prunes most compositions early. The DPO framework provides an inroad to reduce the computational efforts even further. Since each rule is reversible, feasibility can be tested by exploring the space of overall rules from both ends and checking for overlaps at intermediate steps rather then expanding the possible pathways from one end only.

While the focus in this paper is on full rule composition, we illustrate how partial rule composition can be used to automatically detect the required functionality of the catalysts and the additional compounds (in this case a water molecule). Let G' be the graph representation of β-lactam which is the core compound of the reaction. Let $\iota_{G'} = (G', G', G')$ be the identity rule for this compound. The subsequent *partial* composition of the rules, i.e.,

$$\iota_{G'} \circ_{\geq}^c r_1 \circ_{\geq}^c r_2 \circ_{\geq}^c r_3 \circ_{\geq}^c r_4 \circ_{\geq}^c r_5 \circ_{\geq}^c \iota_H$$

result in the overall rule as depicted highlighted in Fig. 7, i.e., any grey molecule or edge disappears. The overall rules show the automatic inference of the necessity of the four functional units of the catalysts and the necessity of the water molecule, as they are subsequently added to the left side of the overall rule during the partial rule composition. When defining graph grammar rules the difficulty often lies in the question of defining the size of the context around a reaction centre: a large context leads to a very specific rule, while a too small context might lead to chemically invalid reactions. Comparing full and partial compositions can be employed as a method to detect the functional units of the catalysts.

The atom mapping of the full composition result shows that in the composed rule with the 8 cycle the acid-base catalysts lysine and glutamic acid are unmodified during the overall process although they are necessary for the mechanism. In the composed rule with the 10 cycle only the acid-base catalyst glutamic acid is unmodified. The other catalysts and the water molecule are modified, however only based on the fact that the hydrogen atom for proton donation is different from the accepting hydrogen.

3.2   Glycolysis

In order to illustrate the potential of rule composition to detect different carbon traces we analyse two variants of the glycolysis pathway. The net reaction of the glycolysis pathway is the conversion of a glucose molecule into two pyruvates while releasing the high-energy compound ATP. For a recent and detailed review see [25]. The most common type of glycolysis is the Embden-Meyerhof-Parnas (EMP) pathway. The alternative Entner-Doudoroff (ED) pathway [26] is known to lead to different carbon atom traces in one of the two pyruvates. Labelling experiments in glycolysis are commonly used to analyse the activity of the different pathways (e.g., [27]). The analysis of such data is quite tedious in practice since the possible atom traces for an overall pathway usually have to be constructed manually.

The individual steps and the enzymes catalysing them are well understood for both the EMP and the ED pathway, and a detailed discussion is far beyond the

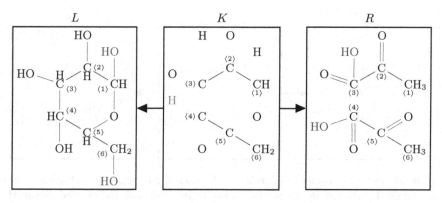

Fig. 8. Simplified transformation rule for the overall EMP pathway with each carbon atom labelled. Some hydroxyl-groups appear to be destroyed and created due to the simplification such that only glucose and pyruvate is depicted.

scope of this paper. The key difference is that EMP yields 2 ATP per glucose while ED produced only 1 ATP for each glucose molecule. In the evolutionary time-line the ED pathway is thought to predate the EMP pathway [28], which is the dominant pathway among eukaryotes. There exists a clear trade-off between the ATP yield and the thermodynamic driving force (rate) of a glycolytic pathway [29]. Flamholz et al. [30] recently found that the high yield EMP pathway needs to maintain much higher enzyme levels (thereby causing greater protein production costs) to support the same flux as through the ED pathway. The two pathways have been modelled using the following transformation rules:

r_1	Pyranose-furanose	r_8	Enolase
r_2	Furanose-linear	r_9	Keto-enol
r_3	Ketose-aldose	r_{10}	NAD+-oxoreductase
r_4	ATP-phosphorylation		
r_5	ATP-dephosphorylation	r_{11}	Lactonohydrolase
r_6	NAD+-phosphorylation	r_{12}	Hydrolyase
r_7	Phosphomutase	r_{13}	Reverse aldolase

The details of the rules can be found in the Appendix of the arXiv version. As in the previous section, we use $\imath_{G(EMP)} = (G, G, G)$ and $\imath_{G(ED)}$ to model input graphs for the two pathways. For EMP G consists of 1 glucose, 2 ATP, 2 ADP, 2 phosphates, and 2 NAD$^+$. In the case of ED the set of input compounds is 1 ATP, 1 ADP, 1 phosphate, and 2 NAD$^+$. $\imath_{H(EMP)} = (H, H, H)$ and $\imath_{H(ED)}$ correspondingly model the output. In the case of EMP the set of output compounds is 2 pyruvates, 4 ATP, 2 NADH, 2 water, and 2 H$^+$. In the case of ED the set of output compounds is 2 pyruvates, 2 ATP, 2 NADH, 2 H$^+$, and 2 water. Note that the same approach as presented in the MACiE 0002 example for automatically inferring the necessary functional groups of G and H could be applied; an explicit definition of the catalysts in G and H would not be necessary.

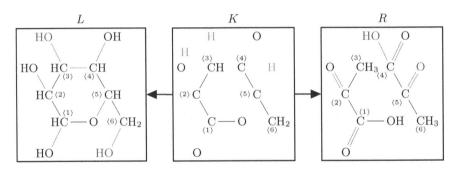

Fig. 9. Simplified transformation rule for the overall ED pathway with each carbon atom labelled. Some hydroxyl-groups appear to be destroyed and created due to the simplification such that only glucose and pyruvate is depicted.

For EMP we compute the composition

$$
\iota_{G(EMP)} \circ \overbrace{\iota_G \circ r_4 \circ r_1 \circ r_4 \circ r_2 \circ r_{13} \circ r_3}^{\text{Glucose} \to 2\ \text{G3P}}
$$
$$
\circ \underbrace{(r_6 \circ_\emptyset r_6) \circ (r_5 \circ_\emptyset r_5) \circ (r_7 \circ_\emptyset r_7) \circ (r_8 \circ_\emptyset r_8) \circ (r_5 \circ_\emptyset r_5) \circ (r_9 \circ_\emptyset r_9)}_{\text{2 G3P} \to 2\ \text{Pyruvate}} \circ \iota_{H(EMP)}
$$

and for ED we compute

$$
\iota_{G(ED)} \circ \underbrace{r_4 \circ r_{10} \circ r_{11} \circ r_{12} \circ r_{13}}_{\text{Glucose} \to \text{G3P} + \text{Pyruvate}} \circ \underbrace{r_6 \circ r_5 \circ r_7 \circ r_8 \circ r_5 \circ r_9}_{\text{G3P} + \text{Pyruvate} \to 2\ \text{Pyruvate}} \circ \iota_{H(ED)}
$$

The resulting rules are depicted in Fig. 8 and Fig. 9. To reduce clutter we only draw the glucose and pyruvate components (formally this can be achieved by composing with rules that unbinds the unwanted components). Clearly, the carbon traces of the two rules differ. Such an approach can be used for an automated design of labelling experiments to detect the activity of pathway alternatives.

The prefixes of the rule composition expression allows to infer all the intermediate compounds and their corresponding atom traces relatively to the input compounds. The summary of this analysis is depicted in Fig. 10 for both pathways (traces shown for carbon atoms only). The black reaction arrows show the EMP pathway, the green arrows show the ED pathway. The six carbon atoms from glucose are converted into two pyruvate molecules in two different ways depending on whether EMP or ED was used to catabolise glucose. While the EMP pathway has a Fructose 1,6-bisphosphate as an intermediate, in which a pentose ring is cleaved, in the ED pathway the hexose ring of the Glucose 6-phosphate is cleaved. The carbon atom trace of one of the two pyruvates is identical, while it is inverted in the other pyruvate.

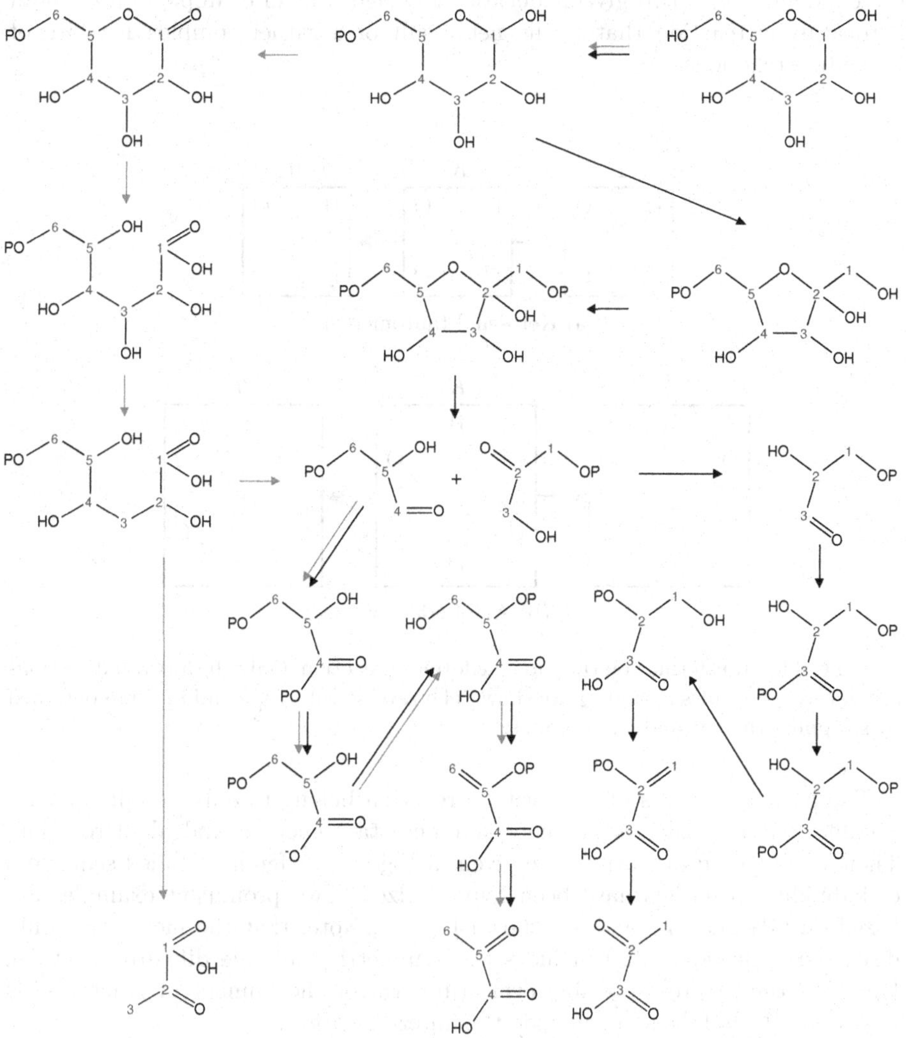

Fig. 10. Carbon atom trace of glycolysis; the Embden-Meyerhof-Parnas pathway (EMP) is depicted with black reaction arrows, the Entner-Doudoroff pathway (ED) is depicted with green reaction arrows. The six carbon atoms from glucose are converted into two pyruvate molecules (highlighted in blue) in two different ways depending on whether EMP or ED was used to catabolise glucose, one pyruvate overlaps in both pathways.

3.3 Formose Reaction

The formose reaction [31] has been extensively discussed as a possible prebiotic route to higher carbohydrates. In contrast to the two previous examples it does not require enzyme catalysis. It converts two formaldeyhydes and a glycolaldehyde into two glycolaldehydes and hence is an example of an overall autocatalytic reaction that is the net result of a rather complex network of individual reactions.

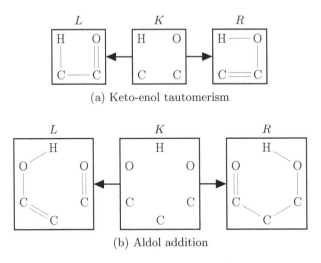

(a) Keto-enol tautomerism

(b) Aldol addition

Fig. 11. The transformation rules for the formose reaction. Only the forward directions of the reversible rules r_1 and r_2 are show. The reverse rules, r_1^{-1} and r_2^{-1}, are obtained by swapping the left and right graphs.

The individual steps of the formose reaction belong to only two distinct reversible reaction patterns, namely keto-enol tautomerism and aldol reaction. Their DPO rule formulations are given in Fig. 11. Different detailed sequences of individual reactions have been hypothesized. Two prominent examples are based on [31] and [32] and are shown Fig. 12. Note, that the most commonly discussed cycle (e.g., [32]) includes the symmetric molecule dihydroxyacetone. However, the shortest possible cycle (in term of the number of reactions) is based on [31] and does not include this intermediate.

The input graph G comprises two formaldehyde and one glycolaldehyde molecule, the goal H consists of two copies of glycolaldehyde. Both are represented by their corresponding identity rules $\iota_G = (G, G, G)$ and $\iota_H = (H, H, H)$. The two proposed pathway are represented as

$$\iota_G \circ r_1 \circ r_2 \circ r_1 \circ r_2 \circ r_1 \circ r_1^{-1} \circ r_2^{-1} \circ r_1^{-1} \circ \iota_H \tag{3}$$

and

$$\iota_G \circ r_1 \circ r_2 \circ r_1 \circ r_1^{-1} \circ r_1 \circ r_2 \circ r_1 \circ r_1^{-1} \circ r_2^{-1} \circ r_1^{-1} \circ \iota_H \tag{4}$$

where r_1, and r_1^{-1} are the keto-enol and enol-keto transitions of keto-enol tautomerism, r_2 is aldol addition and r_2^{-1} is its inverse, i.e., cleavage.

Based on a prefix composition of Eq. (3) and Eq. (4), the traces for the intermediates can be computed as in the glycolysis example. The subsequent

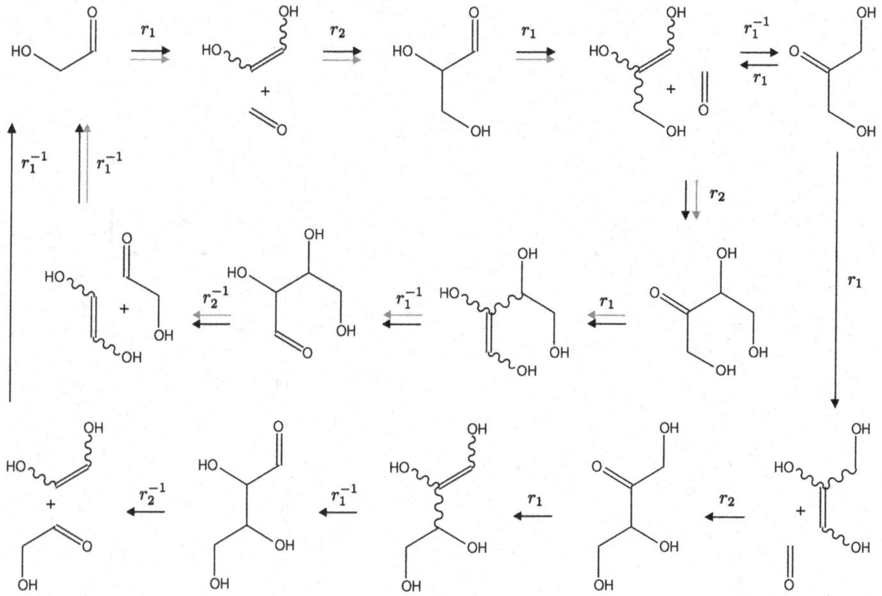

Fig. 12. Detailed mechanism for the formose process. The labels r_1 and r_1^{-1} indicate the keto-enol tautomerisation (forward and backward), r_2 and r_2^{-1} refer to aldol- and retro-aldol reaction. The shortest possible autocatalytic cycle is illustrated with green reaction arrows, the possibilities with dihydroxyacetone as an intermediate are illustrated with black reaction arrows. In order to allow and easy visual tracking of carbon atoms, we duplicated the sequence of compounds after the second aldol addition.

modification of the carbon traces is summarized in Fig. 13. Note that in this figure sequences of isomerisation and aldol-addition steps are depicted as one step in order to minimize clutter.

The rule composition based on Eq. (4) results in six non-isomorphic composed overall rules, each having a different carbon trace for the 4 carbon atoms of G. One of those rules is depicted in Fig. 14. While 4! carbon traces are a trivial upper bound, the mechanism allows only for six of them, as the carbon from the second added formaldehyde cannot end up as carbonyl carbon in the resulting glycolaldehyde. If a labelling experiment could be performed with all carbons uniquely labelled and if the glycolaldehydes after exactly one instantiation of the reaction cycle would be analysed, then not twelve but only nine different glycolaldehydes could be observed. If the mechanism follows Eq. (3) (i.e., dihydroxyacetone is not an intermediate of the mechanism), the two input formaldehydes never combine into the same glycolealdehyde, reducing the set of overall reactions to four rules. Using the same labelling experiment as above, only six different glycolaldehydes could be observed.

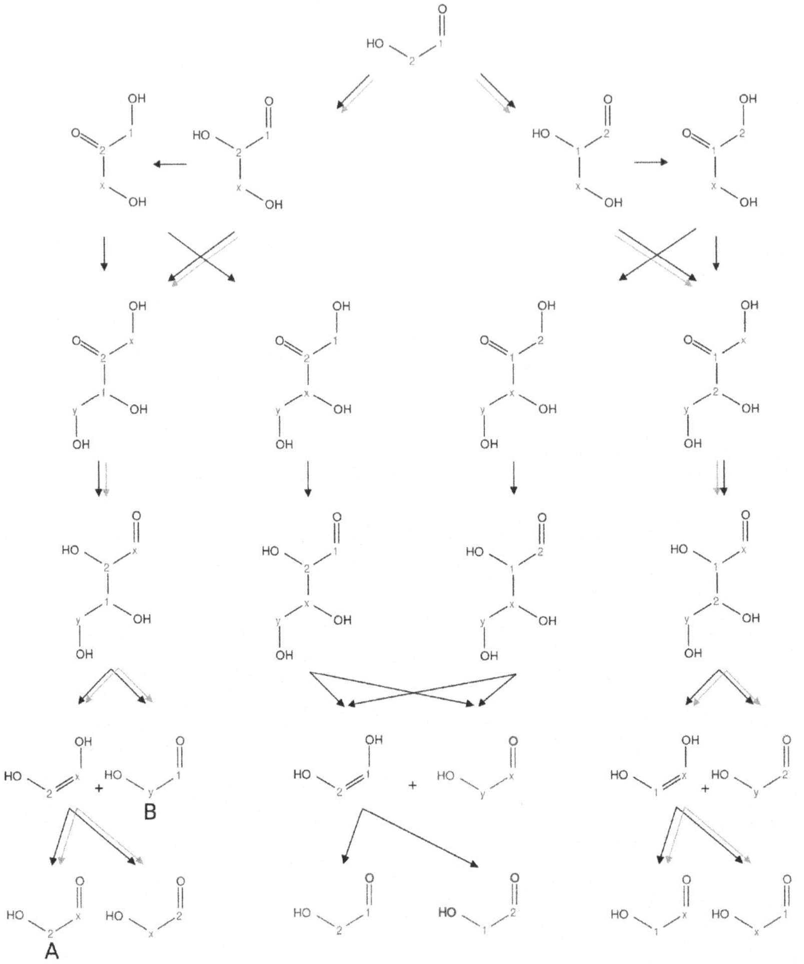

Fig. 13. Carbon atom traces for one round of the formose process: green reaction arrows indicate possible carbon atom traces following the shorter formose cycle, black reaction arrows indicate possible traces following the cycle having dihydroxyacetone as intermediate. The carbonyl (resp. alcohol) carbon of the starting molecule glycolealdehyde is labelled 1 (resp. 2). After condensation of this molecule with two formaldehydes labelled x and y, the intermediate molecule decomposes into two glycolealdehydes. Depending on the mechanism, the labelled carbon atoms end up in nine (resp. six) different positions of the resulting glycolealdehydes (shorter cycle: blue molecules, longer cycle: blue and purple molecules). Note that the carbon from the second formaldehyde (y) can not end up as carbonyl carbon in the resulting glycolealdehydes. The shorter cycle allows for a strict subset of carbon traces only, the two formaldehydes never recombine into a glycolealdehyde. From the six (resp. four) possible composed overall reactions of the longer (resp. shorter) cycle, the one that results in the two blue glycolaldehydes A and B is depicted in Fig. 14.

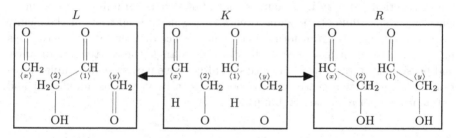

Fig. 14. Formose reaction: one of six possible overall rules based on Eq. (4) including carbon atom trace information. The mapping of the atoms corresponds to creation of the glycolaldehydes denoted A and B in Fig. 13.

4   Conclusions

Chemical reactions are a particularly fruitful area of applications for rule based transformation systems. After all, graphs are the level of abstraction most commonly used by chemists to represent molecules, and the "named reactions" of organic chemistry explicitly are rewriting rules. The composition of multiple individual elementary steps to a single "overall reaction", furthermore, is common practice in the chemistry literature. Rule composition is a powerful tool in graph grammars (and potentially also in other calculi modelling concurrent computation) to investigate such concrete (multi-step) derivations in more detail. This approach is much more general, however, and by no means restricted to the domain of chemistry.

The key idea is to reduce a given derivation to a composite transition rule bound to the initial and final objects. This reduced the problem of representing this composite rule as a sequence of elementary rules from which it is composed. In practical applications there is often only a small number of feasible decompositions that are easily identified by backtracking-style enumeration. DPO graph rewriting offers an interesting advantage for the task at hand. DPO rules are guaranteed to be reversible, hence the search for a path from initial to finite state can be broken up into an exploration from both ends.

Overall chemical reactions are good proving ground for the rule composition approach. Chemical reaction databases usually only list the products and educts. In some cases the reaction mechanisms of the elementary steps catalysed by relevant enzyme(s) are also known, but these lines of information are not connected in a way that would make it easy to retrieve the atom maps. These are key to analysing isotope labelling experiments and hence add practical relevance to our examples. The rule composition framework, however, also provides additional information, such as possible alternative in the relative timing of reaction steps and information on concurrency of elementary reactions. While limited in enzyme reactions (the first two of our three showcase examples), one-pot reactions such as the formose reaction form the other extreme, with large numbers of concurrent reactions. In such a scenario rule composition is a concise way to model reaction pathway in a manner that allows coarse graining away from elementary steps while at the same time allowing us to keep track of chemically distinguishable overall transformations that produce different atom traces. This information is important since it allows to disentangle the relative importance, and possibly even the reaction kinetics, of alternative pathways in complex reaction mixtures. In that sense rule composition can be viewed as an efficient and automatic model reduction technique.

Acknowledgements. This work was supported in part by the *Volkswagen Stiftung*, the COST Action CM1304 "Emergence and Evolution of Complex Chemical Systems", and the Danish Council for Independent Research, Natural Sciences.

References

1. Fontana, W., Buss, L.W.: What would be conserved if "the tape were played twice"? Proc. Natl. Acad. Sci. USA 91, 757–761 (1994)
2. Regev, A., Shapiro, E.: Cells as computation. Nature 419, 343 (2002)
3. Danos, V.: Formal molecular biology. Theor. Comp. Sci. 325, 69–110 (2004)
4. Blinov, M.L., Yang, J., Faeder, J.R., Hlavacek, W.S.: Graph theory for rule-based modeling of biochemical networks. In: Priami, C., Ingólfsdóttir, A., Mishra, B., Riis Nielson, H. (eds.) Transactions on Computational Systems Biology VII. LNCS (LNBI), vol. 4230, pp. 89–106. Springer, Heidelberg (2006)
5. Păun, G.: Computing with membranes. J. Comp. Syst. Sci. 61, 108–143 (2000)
6. Cardelli, L.: Brane calculi. In: Danos, V., Schachter, V. (eds.) CMSB 2004. LNCS (LNBI), vol. 3082, pp. 257–278. Springer, Heidelberg (2005)
7. Arnold, C., Stadler, P.F., Prohaska, S.J.: Chromatin computation: Epigenetic inheritance as a pattern reconstruction problem. J. Theor. Biol. 336, 61–74 (2013)
8. Hlavacek, W., Faeder, J.R., Blinov, M.L., Posner, R.G., Hucka, M., Fontana, W.: Rules for modeling signal-transduction systems. Sci. STKE 6, 334–336 (2006)
9. Sekar, J.A., Faeder, J.R.: Rule-based modeling of signal transduction: a primer. Methods Mol. Biol. 880, 139–218 (2012)
10. Berry, G., Boudol, G.: The chemical abstract machine. In: Proceedings of the 17th ACM SIGPLAN-SIGACT Symposium on Principles of Programming Languages, POPL 1990, pp. 81–94. Assoc. Computing Machinery, New York (1990)
11. Benkö, G., Flamm, C., Stadler, P.F.: A graph-based toy model of chemistry. J. Chem. Inf. Comput. Sci. 43, 1085–1093 (2003)
12. Danos, V., Feret, J., Fontana, W., Harmer, R., Hayman, J., Krivine, J., Thompson-Walsh, C., Winskel, G.: Graphs, rewriting and pathway reconstruction for rule-based models. In: IARCS Annual Conference on Foundations of Software Technology and Theoretical Computer Science (FSTTCS 2012). Leibniz International Proceedings in Informatics (LIPIcs), vol. 18, pp. 276–288 (2012)
13. Beck, M., Benkö, G., Eble, G., Flamm, C., Müller, S., Stadler, P.F.: Graph grammars as models for the evolution of developmental pathways. In: Schaub, H., Detje, F., Brüggemann, U. (eds.) The Logic of Artificial Life: Abstracting and Synthesizing the Principles of Living Systems, pp. 8–15. IOS Press, Akademische Verlagsgesellschaft, Berlin (2004)

14. Sauer, U.: Metabolic networks in motion: ^{13}C-based flux analysis. Mol. Syst. Biol. 2, 62 (2006)

15. Zamboni, N.: ^{13}C metabolic flux analysis in complex systems. Curr. Opin. Biotech. 22, 103–108 (2011)

16. Durot, M., Bourguignon, P.Y., Schachter, V.: Genome-scale models of bacterial metabolisn: reconstruction and applications. FEMS Microbiol. Rev. 33, 164–190 (2009)

17. Feist, A.M., Herrgård, M.J., Thiele, I., Reed, J.L., Palsson, B.Ø.: Reconstruction of biochemical networks in microorganisms. Nature Rev. Microbiol. 7, 129–143 (2009)

18. Holliday, G.L., Bartlett, G.J., Almonacid, D.E., O'Boyle, N.M., Murray-Rust, P., Thornton, J.M., Mitchell, J.B.O.: MACiE: a database of enzyme reaction mechanisms. Bioinformatics 21, 4315–4316 (2005)

19. Holliday, G.L., Andreini, C., Fischer, J.D., Rahman, S.A., Almonacid, D.E., Williams, S.T., Pearson, W.R.: MACiE: exploring the diversity of biochemical reactions. Nucleic Acids Research 40, D783–D789 (2012)

20. Andersen, J.L., Flamm, C., Merkle, D., Stadler, P.F.: Inferring chemical reaction patterns using graph grammar rule composition. J. Syst. Chem. 4(4) (2013)

21. Cook, S.A.: The complexity of theorem-proving procedures. In: Proceedings of the Third Annual ACM Symposium on Theory of Computing, STOC 1971, pp. 151–158. ACM, New York (1971)

22. Weininger, D.: SMILES, a chemical language and information system. 1. Introduction to methodology and encoding rules. J. Chem. Inf. Comput. Sci. 28(1), 31–36 (1988)

23. Weininger, D., Weininger, A., Weininger, J.L.: SMILES 2. Algorithm for Generation of Unique SMILES Notation. J. Chem. Inf. Comput. Sci. 29(2), 97–101 (1989)

24. Atanasov, B.P., Mustafi, D., Makinen, M.W.: Protonation of the beta-lactam nitrogen is the trigger event in the catalytic action of class A beta-lactamases. Proc. Natl. Acad. Sci. 97(7), 3160–3165 (2000)

25. Bar-Even, A., Flamholz, A., Noor, E., Milo, R.: Rethinking glycolysis: on the biochemical logic of metabolic pathways. Nat. Chem. Biol. 8(6), 509–517 (2012)

26. Entner, N., Doudoroff, M.: Glucose and gluconic acid oxidation of pseudomonas saccharophila. J. Biol. Chem. 196, 853–862 (1952)

27. Borodina, I., Schöller, C., Eliasson, A., Nielsen, J.: Metabolic network analysis of streptomyces tenebrarius, a streptomyces species with an active entner-doudoroff pathway. Appl. Environ. Microbiol. 71(5), 2294–2302 (2005)

28. Romano, A.H., Conway, T.: Evolution of carbohydrate metabolic pathways. Res. Microbiol. 147(6/7), 448–455 (1996)

29. Stettner, A.I., Segré, D.: The cost of efficiency in energy metabolism. PNAS 110(24), 9629–9630 (2013)

30. Flamholz, A., Noor, E., Bar-Even, A., Liebmeister, W., Milo, R.: Glycolytic stratewgy as a tradeoff between energy yield and protein cost. PNAS 110(24), 10039–10044 (2013)

31. Benner, S., Kim, H., Ricardo, A.: Planetary organic chemistry and the origins of biomolecules. Cold Spring Harb. Perspect. Biol. 2(7), a003467 (2010)

32. Breslow, R.: On the mechanism of the formose reaction. Tetrahedron Letters 1(21) (1959)

A Logical Framework for Systems Biology

Elisabetta de Maria[1], Joëlle Despeyroux[2], and Amy P. Felty[3]

[1] Laboratoire I3S, University of Nice - Sophia-Antipolis, Sophia-Antipolis, France
edemaria@i3s.unice.fr
[2] INRIA and CNRS, Laboratoire I3S, UNS, Sophia-Antipolis, France
joelle.despeyroux@inria.fr
[3] School of Electrical Engineering and Computer Science,
University of Ottawa Ottawa, Canada
afelty@eecs.uottawa.ca

Abstract. We propose a novel approach for the formal verification of biological systems based on the use of a modal linear logic. We show how such a logic can be used, with worlds as instants of time, as an unified framework to encode both biological systems and temporal properties of their dynamic behaviour. To illustrate our methodology, we consider a model of the P53/Mdm2 DNA-damage repair mechanism. We prove several properties that are important for such a model to satisfy and serve to illustrate the promise of our approach. We formalize the proofs of these properties in the Coq Proof Assistant, with the help of a Lambda Prolog prover for partial automation of the proofs.

1 Introduction

In this paper, we consider the question of reasoning about biological systems in a modal linear logic. We show that a new logic, called Hybrid Linear Logic (HyLL) developed by the second author in joint work with K. Chaudhuri [8,14], is particularly well-suited to this purpose. HyLL provides a unified framework to encode biological systems, to express temporal properties of their dynamic behaviour, and to prove these properties. By constructing proofs in the HyLL logic, we directly witness reachability as logical entailment. This approach is in contrast to most current approaches to applying formal methods to systems biology, which generally encode biological systems either in a dedicated programming language or in differential equations, express properties in a temporal logic, and then verify these properties against some form of traces built using an external simulator. In the next subsection, we review in some detail the state of the art of such approaches, in order to further situate and motivate our new approach. In Sect. 1.2, we motivate our choice of linear logic in general and HyLL in particular. Then in Sect. 1.3, we further outline our contributions as well as the overall organization of the rest of the paper.

1.1 Formal Methods for Systems Biology

Computational systems biology provides a variety of methods for understanding the structure of biological systems and for studying their dynamics, that is, the temporal evolution of the involved entities.

F. Fages and C. Piazza (Eds.): FMMB 2014, LNBI 8738, pp. 136–155, 2014.

To capture the qualitative nature of dynamics, Thomas introduced a Boolean approach for regulatory networks (an entity is present or absent) [32] and subsequently generalized it to multivalued levels of concentration [33]. Contrary to Petri nets, which are based on synchronous updating techniques [29], Thomas' discrete models are asynchronous. Other purely qualitative approaches are π-calculus [31], bio-ambients [30], and reaction rules [13,7].

To describe the dynamics from a quantitative point of view, ordinary or stochastic differential equations are heavily used. More recent approaches include hybrid Petri nets [22] and hybrid automata [1,5], piecewise linear equations [24], stochastic π-calculus [28], and rule-based languages with continuous/stochastic dynamics such as Kappa [13].

The Biochemical Abstract Machine Biocham [15] is a framework that allows the description of a biochemical system in terms of reaction rules and the interpretation of it at different levels of abstraction, by either an asynchronous Boolean transition system (Boolean semantics), a continuous time Markov chain (stochastic semantics), or a system of ordinary differential equations over molecular concentrations (differential semantics).

One of the most common approaches to the formal verification of biological systems is model checking [11]. Model checking allows one to verify desirable properties of a system by an exhaustive enumeration of all the states reachable by the system. In order to apply such a technique, the biological system should be encoded as a finite transition system and relevant system properties should be specified using propositional temporal logic. Formally, a transition system over a set AP of atomic propositions is a tuple $M = (Q, T, L)$, where Q is a finite set of states, $T \subseteq Q \times Q$ is a total transition relation (that is, for every state $q \in Q$ there is a state $q' \in Q$ such that $T(q, q')$), and $L : Q \to 2^{AP}$ is a labeling function that maps every state into the set of atomic propositions that hold at that state.

Temporal logics are formalisms for describing sequences of transitions between states. The computation tree logic CTL* allows one to describe properties of computation trees. Its formulas are obtained by (repeatedly) applying Boolean connectives, *path quantifiers*, and *state quantifiers* to atomic formulas. The path quantifier A (resp., E) can be used to state that all paths (resp., some path) starting from a given state have some property. The state quantifiers are X (next time), F (sometimes in the future), G (always in the future), and U (until). The branching time logic CTL is a fragment of CTL* that allows quantification over the paths starting from a given state. Unlike CTL*, it constrains every state quantifier to be immediately preceded by a path quantifier. The linear time logic LTL is another known fragment of CTL* where one may only describe events along a single computation path. The Probabilistic Computation Tree Logic PCTL quantifies the different paths by replacing the E and A modalities of CTL by probabilities.

In Biocham, CTL, LTL, and a fragment of PCTL with numerical constraints are used in the three semantics of reaction models, respectively, in the boolean semantics, in the differential semantics and in the stochastic semantics.

Given a transition system $M = (Q, T, L)$, a state $q \in Q$, and a temporal logic formula φ expressing some desirable property of the system, the *model checking problem* consists of establishing whether φ holds at q or not, namely, whether $M, q \models \varphi$. Another formulation of the model checking problem consists of finding all the states $q \in Q$ such that $M, q \models \varphi$. Observe that the second formulation is more general than the first one.

There exist several tools for checking if a finite state system verifies a given CTL, LTL, or PCTL formula, e.g., NuSMV [10], SPIN [23], and PRISM [20].

In contrast to the above approaches, in our new technique we encode both biological systems and temporal properties in HyLL, and prove that the properties can be derived from the system. We focus on Boolean systems and in this case a time unit corresponds to a transition in the system. We believe that discrete modeling is crucial in systems biology because it allows taking into account some phenomena that have a very low chance of happening (and could thus be neglected by differential approaches), but which may have a strong impact on system behaviour.

1.2   Linear Logic

Linear Logic (LL) [19] is particularly well suited for describing state transition systems. LL has been successfully used to model such diverse systems as: Petri nets, CCS, the π-calculus [6,26], concurrent ML [6], security protocols [4], multiset rewriting, and games.

In the area of biology, for example, a rule of activation (e.g., a protein activates a gene or the transcription of another protein) can be modeled by the following LL axiom:

$$\texttt{active}(a, b) \stackrel{\text{def}}{=} \texttt{pres}(a) \rightarrow (\texttt{pres}(a) \otimes \texttt{pres}(b)).$$

The formula $\texttt{active}(a, b)$ describes the fact that a state where a is present ($\texttt{pres}(a)$ is true) can evolve into a state where both $\texttt{pres}(a)$ and $\texttt{pres}(b)$ are true.

Propositions such as $\texttt{pres}(a)$ are called *resources*, and a rule in the logic can be viewed as a rewrite rule from a set of resources into another set of resources, where a set of resources describes a state of the system. Thus, a particular state transition system can be modeled by a set of rules of the above shape. The rules of the logic then allow us to prove some desired properties of the system, such as, for example, the existence of a stable state.

However, linear implication is timeless: there is no way to correlate two concurrent transitions. If resources have lifetimes and state changes have temporal, probabilistic or stochastic *constraints*, then the logic will allow inferences that may not be realizable in the system being modeled. This was the motivation of the development of HyLL, which was designed to represent constrained transition systems.

1.3   Contributions and Organization

In this work, we present some first applications of HyLL to systems biology. We present HyLL in Sect. 2 and the overall approach to the application domain in Sect. 3. We choose a simple yet representative biological example concerning the DNA-damage repair mechanism based on proteins p53 and Mdm2, and present and prove several properties of this system (Sect. 4). We fully formalize these proofs in a theorem prover we have implemented in the Coq Proof Assistant [2] and λProlog [27] (Sect. 5). This prover is designed to both reason in HyLL and to formalize meta-theoretic properties about it.

We discuss the merits and eventual drawbacks of this new approach compared to approaches using temporal logic and model checking. To better illustrate the correspondence with such approaches, we also present in some detail the encoding of temporal logic operators in HyLL (Sect. 6).

We conclude and discuss future work in Sect. 7.

An electronic appendix (`fmmb-eappendix.pdf`) as well as the formal proofs and a report describing our informal proofs in more detail can be found at `www.eecs.uottawa.ca/~afelty/fmmb14/`. All references to the appendix in this paper point to this web site.

2   A Hybrid Linear Logic

HyLL is a conservative extension of intuitionistic first-order linear logic (LL) [19] where the truth judgements are parameterized on a *constraint domain*: $A @ w$ stands for the truth of A under constraint w. A typical example of such a judgement is "A is true at time t", or "with probability p."

2.1   HyLL Syntax

Like in the linear logic LL, propositions are interpreted as *resources* which may be composed into a *state* using the usual linear connectives, and the linear implication ($\rightarrow$) denotes a transition between states. The world label w of a judgement $A @ w$ represents a constraint on states and state transitions; particular choices for the worlds produce particular instances of HyLL. The common component in all the instances of HyLL is the proof theory, which is fixed once and for all. The minimal requirement on the kinds of constraints that HyLL can deal with is defined as follows:

Definition 1. *A* constraint domain $\mathcal{W}$ *is a monoid structure* $\langle W, ., \iota \rangle$. *The elements of* W *are called* worlds, *and the partial order* $\preceq : W \times W$ —*defined as* $u \preceq w$ *if there exists* $v \in W$ *such that* $u.v = w$—*is the* reachability relation *in* $\mathcal{W}$.

The identity world ι is $\preceq$-initial and is intended to represent the lack of any constraints. Thus, the ordinary first-order linear logic is embeddable into any instance of HyLL by setting all world labels to the identity. A typical and simple example of constraint domain is $\mathcal{T} = \langle \Re^+, +, 0 \rangle$, or $\langle I\!N, +, 0 \rangle$, representing instants of time.

Atomic propositions are written using lowercase letters $(p, q, ...)$ applied to a sequence of *terms* $(s, t, ...)$, which are drawn from an untyped term language containing constants $(c, ...)$, term variables $(x, y, ...)$ and function symbols $(f, g, ...)$ applied to a list of terms $(\vec{t})$. Non-atomic propositions are constructed from the connectives of first-order intuitionistic linear logic and the two hybrid connectives *satisfaction* (at), which states that a proposition is true at a given world $(w, \iota, u.v, ...)$, and *localization* $(\downarrow)$, which binds a name for the (current) world the proposition is true at. The following grammar summarizes the syntax of HyLL terms (t), and propositions (A, B).

$$t \quad ::= c \mid x \mid f(\vec{t})$$
$$A, B ::= p(\vec{t}) \mid A \otimes B \mid \mathbf{1} \mid A \to B \mid A \,\&\, B \mid \top \mid A \oplus B \mid \mathbf{0} \mid !A \mid$$
$$\forall x.\ A \mid \exists x.\ A \mid (A \text{ at } w) \mid \downarrow u.\ A \mid \forall u.\ A \mid \exists u.\ A$$

Note that world u is bound in the propositions $\downarrow u.A$, $\forall u.\ A$ and $\exists u.\ A$.

World variables cannot be used in terms, and neither can term variables occur in worlds; this restriction is important for the modular design of HyLL because it keeps purely logical truth separate from constraint truth. We let α range over variables of either kind. Note that $\downarrow$ and at commute freely with all non-hybrid connectives [8].

2.2   Sequent Calculus for HyLL

We present the syntax of hybrid logic in a sequent calculus style [18], using sequents of the form $\Gamma; \Delta \vdash C @ w$ where Γ and Δ are sets of judgements of the form $A @ w$, with Δ being moreover a *multiset*. Γ is called the *unrestricted context*: its hypotheses can be consumed any number of times. Δ is a *linear context*: every hypothesis in it must be consumed singly in the proof. Note that in a judgement $A @ w$ (as in a proposition A at w), w can be any expression in $\mathcal{W}$, not only a variable.

The main set of inference rules is in Fig. 1 (the complete set is in Appendix A). The rules for the first-order quantifiers (omitted) and the exponential ! are completely standard. A brief discussion of the hybrid rules follows. To introduce the *satisfaction* proposition $(A$ at $u)$ (at any world w) on the right, the proposition A must be true in the world u. The proposition $(A$ at $u)$ itself is then true at any world, not just in the world u. In other words, $(A$ at $u)$ carries with it the world at which it is true. Therefore, suppose we know that $(A$ at $u)$ is true (at any world v); then, we also know that $A @ u$, and we can use this hypothesis (rule "at L"). The other hybrid connective of *localisation*, $\downarrow$, is intended to be able to name the current world. That is, if $\downarrow u.\ A$ is true at world w, then the variable u stands for w in the body A. This interpretation is reflected in its right introduction rule $\downarrow R$. For left introduction, suppose we have a proof of $\downarrow u.\ A @ v$ for some world v. Then, we also know, and thus can use $A[v/u] @ v$.

Note that there are only two structural rules in HyLL. Weakening and contraction are admissible rules. For example:

Judgemental rules

$$\Gamma; p(\vec{t}) \;@\; w \vdash p(\vec{t}) \;@\; w \;[init] \qquad \frac{\Gamma, A \;@\; u; \Delta, A \;@\; u \vdash C \;@\; w}{\Gamma, A \;@\; u; \Delta \vdash C \;@\; w} \; copy$$

Multiplicative

$$\frac{\Gamma; \Delta \vdash A \;@\; w \quad \Gamma; \Delta' \vdash B \;@\; w}{\Gamma; \Delta, \Delta' \vdash A \otimes B \;@\; w} \otimes R \qquad \frac{\Gamma; \Delta, A \;@\; u, B \;@\; u \vdash C \;@\; w}{\Gamma; \Delta, A \otimes B \;@\; u \vdash C \;@\; w} \otimes L$$

$$\Gamma; . \vdash \mathbf{1} \;@\; w \;[1R] \qquad \frac{\Gamma; \Delta \vdash C \;@\; w}{\Gamma; \Delta, \mathbf{1} \;@\; u \vdash C \;@\; w} \; 1\,L$$

$$\frac{\Gamma; \Delta, A \;@\; w \vdash B \;@\; w}{\Gamma; \Delta \vdash A \to B \;@\; w} \to R \qquad \frac{\Gamma; \Delta \vdash A \;@\; u \quad \Gamma; \Delta', B \;@\; u \vdash C \;@\; w}{\Gamma; \Delta, \Delta', A \to B \;@\; u \vdash C \;@\; w} \to L$$

Additive

$$\Gamma; \Delta \vdash \top \;@\; w \;[\top R] \qquad \Gamma; \Delta, \mathbf{0} \;@\; u \vdash C \;@\; w \;[\mathbf{0}L]$$

$$\frac{\Gamma; \Delta \vdash A \;@\; w \quad \Gamma; \Delta \vdash B \;@\; w}{\Gamma; \Delta \vdash A \,\&\, B \;@\; w} \,\&\, R \qquad \frac{\Gamma; \Delta, A_i \;@\; u \vdash C \;@\; w}{\Gamma; \Delta, A_1 \,\&\, A_2 \;@\; u \vdash C \;@\; w} \,\&\, L_i$$

$$\frac{\Gamma; \Delta \vdash A_i \;@\; w}{\Gamma; \Delta \vdash A_1 \oplus A_2 \;@\; w} \oplus R_i \qquad \frac{\Gamma; \Delta, A \;@\; u \vdash C \;@\; w \quad \Gamma; \Delta, B \;@\; u \vdash C \;@\; w}{\Gamma; \Delta, A \oplus B \;@\; u \vdash C \;@\; w} \oplus L$$

Exponentials rules

$$\frac{\Gamma; . \vdash A \;@\; w}{\Gamma; . \vdash \,!A \;@\; w} \;!R \qquad \frac{\Gamma, A \;@\; u; \Delta \vdash C \;@\; w}{\Gamma; \Delta, !A \;@\; u \vdash C \;@\; w} \;!L$$

Hybrid connectives

$$\frac{\Gamma; \Delta \vdash A \;@\; u}{\Gamma; \Delta \vdash (A \text{ at } u) \;@\; w} \;[\text{at } R] \qquad \frac{\Gamma; \Delta, A \;@\; u \vdash C \;@\; w}{\Gamma; \Delta, (A \text{ at } u) \;@\; v \vdash C \;@\; w} \;[\text{at } L]$$

$$\frac{\Gamma; \Delta \vdash A[w/u] \;@\; w}{\Gamma; \Delta \vdash \downarrow u.A \;@\; w} \;[\downarrow R] \qquad \frac{\Gamma; \Delta, A[v/u] \;@\; v \vdash C \;@\; w}{\Gamma; \Delta, \downarrow u.A \;@\; v \vdash C \;@\; w} \;[\downarrow L]$$

Fig. 1. (Part of) the sequent calculus for HyLL

Theorem 2 (weakening). *If* $\Gamma; \Delta \vdash C \;@\; w$, *then* $\Gamma, \Gamma'; \Delta \vdash C \;@\; w$.

The most important structural properties are the admissibility of the identity and cut theorem; the latter guarantees consistency.

Theorem 3 (identity). $\Gamma, A \;@\; w \vdash A \;@\; w$.

Theorem 4 (cut)
1. *If* $\Gamma; \Delta \vdash A \;@\; u$ *and* $\Gamma; \Delta', A \;@\; u \vdash C \;@\; w$, *then* $\Gamma; \Delta, \Delta' \vdash C \;@\; w$
2. *If* $\Gamma; . \vdash A \;@\; u$ *and* $\Gamma, A \;@\; u; \Delta \vdash C \;@\; w$, *then* $\Gamma; \Delta \vdash C \;@\; w$.

An example of derived statements, true in every semantics for worlds, is the following:

Proposition 5 (relocalisation). *Any true judgement can be relocated at any time in the future:*

$$\frac{\Gamma; A_1 @ w_1 \cdots A_k @ w_k \vdash B @ v}{\Gamma; A_1 @ u.w_1 \cdots A_k @ u.w_k \vdash B @ u.v}$$

This property is particularly well suited to applications in biology. The interested reader can find proofs and further meta-theoretical theorems about HyLL in [8].

2.3   Some Definitions for Biology

We can define modal connectives in HyLL as follows:

Definition 6 (modal connectives)

$$\Box A \overset{\text{def}}{=} \downarrow u. \forall w. (A \ \text{\textit{at}} \ u.w) \qquad \Diamond A \overset{\text{def}}{=} \downarrow u. \exists w. (A \ \text{\textit{at}} \ u.w)$$
$$\delta_v A \overset{\text{def}}{=} \downarrow u. (A \ \text{\textit{at}} \ u.v) \qquad \dagger A \overset{\text{def}}{=} \forall u. (A \ \text{\textit{at}} \ u)$$

The connective δ represents a form of delay. Note its derived right rule:

$$\frac{\Gamma \vdash A @ w.v}{\Gamma \vdash \delta_v A @ w} \ [\delta R]$$

The proposition $\delta_v A$ thus stands for an *intermediate state* in a transition to A. Informally it can be thought to be "v before A". The modally unrestricted proposition $\dagger A$ represents a resource that is consumable in any world; it is mainly used to make transition rules applicable at all worlds.

Oscillation is one of the typical properties of interest in biological systems (illustrated here by Property 1 in Sect. 4.3). In our logic, we can define one oscillation between A and B, with respective delays u and v, as follows:

Definition 7 (one oscillation)
$$oscillate_1 (A, B, u, v) \overset{\text{def}}{=} A \ \& \ \delta_u(B \ \& \ \delta_v A) \ \& \ (A \& B \to 0)$$

Note that the above HyLL proposition closely corresponds to the temporal formula $A \wedge \mathsf{EF}(B \wedge \mathsf{EF}A)$.

Oscillation can be more generally defined by the following proposition in HyLL:

Definition 8 (oscillation)
$$oscillate_h (A, B, u, v) \overset{\text{def}}{=} \dagger[(A \to \delta_u B) \ \& \ (B \to \delta_v A)] \ \& \ (A \& B \to 0).$$

However, since oscillation can be considered a meta-level property of the biological systems modeled in HyLL, this property is perhaps more naturally defined as follows:

Definition 9 (oscillation). $oscillate (A, B, u, v) \overset{\text{def}}{=}$ *for any w,*
$(A @ w \vdash B @ w.u)$, $(B @ w.u \vdash A @ w.u.v)$, *and* $(\vdash A \& B \to 0 @ w)$.

2.4   Temporal Constraints

In this paper, we only consider the constraint domain $\mathcal{T} = \langle I\!N, +, 0 \rangle$ representing (discrete) instants of time, and we write HyLL[$\mathcal{T}$] for this instantiation of HyLL. Delay (Definition 6) in HyLL[$\mathcal{T}$] represents intervals of time; $\delta_d A$ means "A will become available after delay d". For our temporal specifications the notion of addition on times is fundamental.

3   Approach

In this work we take into consideration Boolean models consisting of (i) a set of Boolean variables, (ii) a (partially defined) initial state denoting the presence/absence of (some) variables, and (iii) a set of rules of the form $L_i \Rightarrow R_i$, where the left (resp. right) hand side of the rule L_i (resp. R_i) is the conjunction of a set of predicates concerning the presence/absence of variables. For example, $x_1 \wedge x_2 \Rightarrow \neg x_3$ is a valid rule. This kind of rule can be used to describe state transitions involving control variables or abstract processes. In Biocham [15], they are mostly used to represent biochemical reactions (observe that Biocham rules are more restrictive, they do not express the absence of a variable). In this paper, we take advantage of these rules to express influence rules (e.g. activations and inhibitions) in a biological system.

Observe that, given a Boolean model of this kind, although we do not do it, it is always possible to build a transition system where the set of states is the set of all tuples of Boolean values denoting the presence/absence of the different variables, and a pair of states (s, s') belongs to the relation if and only if there exists a rule i such that s satisfies L_i, s' satisfies R_i, and all the variables not involved in rule i have the same value in s and s'.

If L_i speaks about the presence of a variable x, nothing can be said about the presence/absence of x in s'. In other words, nothing can be inferred about the consumption of variables appearing in the left hand side of rules. If we want to force consumption/not consumption, we have to specify it explicitly. Furthermore, our rules are asynchronous: one rule can be fired at a time. If several rules can be taken from a given state, one of them is non-deterministically chosen. As in Biocham, we choose an asynchronous semantics in order to eliminate the risk of affecting fundamental biological phenomena such as the masking of a relation by another one and the consequent inhibition/activation of biological processes.

To verify whether a Boolean model satisfies a given temporal property, our approach consists of encoding both the model and the property in the HyLL logic and producing a proof. Observe that we do not explicitly build the transition system, we just give a set of variables and rules. Proving temporal properties can result in building a sub-part of the system. There is an analogy with on-the-fly model checking [12], a technique that in many cases avoids the construction of the entire state space of the system (because the property to test guides the construction of the system).

4   Example

In this section we focus on the P53/Mdm2 DNA-damage repair mechanism. P53 is a tumor suppressor protein that is activated in reply to DNA damage. In normal conditions, the concentration of p53 in the nucleus of a cell is weak: its level is controlled by another protein, Mdm2. These two proteins present a loop of negative regulation. In fact, P53 activates the transcription of Mdm2 while the latter accelerates the degradation of the former.

DNA damage increases the degradation rate of Mdm2 so that the control of this protein on P53 becomes weaker and the concentration of p53 can increase. P53 can thus exercise its functions, either stopping the cell cycle to allow DNA repair, or provoking apoptosis, if damage is too heavy.

When Mdm2 loosens its influence on P53, it is possible to observe some oscillations of P53 and Mdm2 concentrations. The answer to a stronger damage is a bigger number of oscillations. In the literature, several models have been proposed to model the oscillatory behaviour of proteins P53 and Mdm2 (see e.g. Ciliberto et al. [9]). In [25] it is shown how a model of the P53/Mdm2 DNA-damage repair mechanism can be exploited in the study of cancer therapies.

4.1   Definition

In the following we propose a simple Boolean model considering the presence/absence of the three variables DNAdam, P53, and MdM2. Initial states are the ones where P53 is absent and Mdm2 is present.

The behaviour of the biological system is specified by the six following rules:

$$
\begin{array}{ll}
\text{1) Dnadam} \Rightarrow \neg\text{Mdm2} & \text{4) Mdm2} \Rightarrow \neg\text{P53} \\
\text{2) } \neg\text{Mdm2} \Rightarrow \text{P53} & \text{5) P53} \Rightarrow_C \neg\text{Dnadam} \\
\text{3) P53} \Rightarrow \text{Mdm2} & \text{6) } \neg\text{Dnadam} \Rightarrow \text{Mdm2}
\end{array}
$$

In rule 5, we use $\Rightarrow_C$ to force consumption, i.e., if P53 is present, after firing the rule both P53 and Dnadam are absent. Note that, if Dnadam is present in the initial state, this rule (that refers to damage repair) can be non-deterministically fired after one, a few, or several P53/MdM2 oscillations. This is consistent with experiments, where the number of oscillations preceding damage repair depends on the damage entity. In the other rules we assume there is no consumption.

4.2   Specification in HyLL

The biological system is modeled in HyLL by a set of axioms of two kinds. First, each rule of the biological system is modeled by a formula in HyLL, as it would be in ordinary linear logic, with the additional use of the delay operator δ_v, making precise the delay taken by the corresponding transition. The domain of world is $\mathcal{T} = \langle \mathbb{N}, +, 0 \rangle$ and we fix all time delays to 1. Then we add a set of axioms stating some well-definedness conditions and the initial state. To encode the basic Boolean model, we use two predicates: $\texttt{pres}(a)$ (seen earlier) and $\texttt{abs}(a)$ to indicate the presence or absence of variable a. The full model is given in Fig. 2.

Activation/Inhibition Rules. For the sake of clarity, we first define generic activation/inhibition actions. These actions can be defined in various ways. A first attempt at describing an activation rule without consumption, for example, might be the following:

$$w\_active(a, b) \stackrel{\text{def}}{=} \mathtt{pres}(a) \to \delta_1(\mathtt{pres}(a) \otimes \mathtt{pres}(b)).$$

Note that in the case where b is present, the above rule does not modify the state. In order to avoid uninteresting proofs, we might alternatively define our activation rule in a more precise way, as follows:

$$s\_active(a, b) \stackrel{\text{def}}{=} \mathtt{pres}(a) \otimes \mathtt{abs}(b) \to \delta_1(\mathtt{pres}(a) \otimes \mathtt{pres}(b)).$$

We call the first kind of rules *weak* rules, and the second one *strong* rules. For the sake of completeness, let us mention a third kind of rules, that we might call *useless*:

$$u\_active(a, b) \stackrel{\text{def}}{=} \mathtt{pres}(a) \otimes \mathtt{pres}(b) \to \delta_1(\mathtt{pres}(a) \otimes \mathtt{pres}(b)).$$

The *general* form of an activation rule, taking in account the various possible values of b, and our eventual missing knowledge of this, is then the following:

$$active(a, b) \stackrel{\text{def}}{=} (\mathtt{pres}(a) \oplus (\mathtt{pres}(a) \otimes \mathtt{pres}(b)) \oplus (\mathtt{pres}(a) \otimes \mathtt{abs}(b))) \\ \to \delta_1(\mathtt{pres}(a) \otimes \mathtt{pres}(b)).$$

The properties stated in the present paper will all use the general form of the biological rules, except Property 4, which is only valid for strong rules.

There are further alternatives for the activation/inhibition rules. Let us consider the above strong variant $s\_active(a, b)$. We can define an *activation with consumption* (of the product a) as follows:

$$s\_active_c(a, b) \stackrel{\text{def}}{=} \mathtt{pres}(a) \otimes \mathtt{abs}(b) \to \delta_1(\mathtt{abs}(a) \otimes \mathtt{pres}(b)),$$

while a *strong activation* will have an inhibitor effect, in case of absence of a:

$$s\_active_s(a, b) \stackrel{\text{def}}{=} \mathtt{abs}(a) \otimes \mathtt{pres}(b) \to \delta_1(\mathtt{abs}(a) \otimes \mathtt{abs}(b)).$$

Our example also uses the corresponding three kinds of rules for the inhibition actions (see Fig. 2). Of course, we could also define activation/inhibition rules accounting for a lack of information concerning consumption.

The System. Before giving the complete definition of our system, we need to additionally specify, in each rule, that if a variable is not touched, then its value remains the same in the next state. This is the purpose of the **unchanged** predicate, which in turn leads us to introduce a parameter **vars**, specifying the set of variables of the biological system.

Note that the definition of the **unchanged** predicate relies on the hypothesis (discussed earlier, in Sect. 3) that only one rule of the biological system can fire at a time. Note also the use of the ! and & operators in its definition: this is an intuitionistic predicate.

- *Variables*:
  $$\mathtt{unchanged}(x, w) \stackrel{\mathrm{def}}{=}\ ! \,[(\mathtt{pres}(x) \text{ at } w \to \mathtt{pres}(x) \text{ at } w.1)\ \&$$
  $$(\mathtt{abs}(x) \text{ at } w \to \mathtt{abs}(x) \text{ at } w.1)].$$
  $$\mathtt{unchanged}(V, w) \stackrel{\mathrm{def}}{=} \otimes_{x \in V} \mathtt{unchanged}(x, w).$$
- *Activation*:
  $$\mathtt{active}(V, a, b) \stackrel{\mathrm{def}}{=} (\mathtt{pres}(a) \oplus (\mathtt{pres}(a) \otimes \mathtt{pres}(b)) \oplus (\mathtt{pres}(a) \otimes \mathtt{abs}(b)))$$
  $$\to \delta_1(\mathtt{pres}(a) \otimes \mathtt{pres}(b)) \otimes \downarrow u.\ \mathtt{unchanged}(V \setminus \{a, b\}, u)).$$
- *Activation with consumption*:
  $$\mathtt{active}_c(V, a, b) \stackrel{\mathrm{def}}{=} (\mathtt{pres}(a) \oplus (\mathtt{pres}(a) \otimes \mathtt{pres}(b)) \oplus (\mathtt{pres}(a) \otimes \mathtt{abs}(b)))$$
  $$\to \delta_1(\mathtt{abs}(a) \otimes \mathtt{pres}(b)) \otimes \downarrow u.\ \mathtt{unchanged}(V \setminus \{a, b\}, u)).$$
- *Strong activation*:
  $$\mathtt{active}_s(V, a, b) \stackrel{\mathrm{def}}{=} (\mathtt{abs}(a) \oplus (\mathtt{abs}(a) \otimes \mathtt{pres}(b)) \oplus (\mathtt{abs}(a) \otimes \mathtt{abs}(b)))$$
  $$\to \delta_1(\mathtt{abs}(a) \otimes \mathtt{abs}(b)) \otimes \downarrow u.\ \mathtt{unchanged}(V \setminus \{a, b\}, u)).$$
- *Inhibition*:
  $$\mathtt{inhib}(V, a, b) \stackrel{\mathrm{def}}{=} (\mathtt{pres}(a) \oplus (\mathtt{pres}(a) \otimes \mathtt{pres}(b)) \oplus (\mathtt{pres}(a) \otimes \mathtt{abs}(b)))$$
  $$\to \delta_1(\mathtt{pres}(a) \otimes \mathtt{abs}(b)) \otimes \downarrow u.\ \mathtt{unchanged}(V \setminus \{a, b\}, u)).$$
- *Inhibition with consumption*:
  $$\mathtt{inhib}_c(V, a, b) \stackrel{\mathrm{def}}{=} (\mathtt{pres}(a) \oplus (\mathtt{pres}(a) \otimes \mathtt{pres}(b)) \oplus (\mathtt{pres}(a) \otimes \mathtt{abs}(b)))$$
  $$\to \delta_1(\mathtt{abs}(a) \otimes \mathtt{abs}(b)) \otimes \downarrow u.\ \mathtt{unchanged}(V \setminus \{a, b\}, u)).$$
- *Strong inhibition*:
  $$\mathtt{inhib}_s(V, a, b) \stackrel{\mathrm{def}}{=} (\mathtt{abs}(a) \oplus (\mathtt{abs}(a) \otimes \mathtt{pres}(b)) \oplus (\mathtt{abs}(a) \otimes \mathtt{abs}(b)))$$
  $$\to \delta_1(\mathtt{abs}(a) \otimes \mathtt{pres}(b)) \otimes \downarrow u.\ \mathtt{unchanged}(V \setminus \{a, b\}, u)).$$
- *Well definedness*:
  $$\mathtt{well\_defined}_0(V) \stackrel{\mathrm{def}}{=} \forall a \in V.\ [\mathtt{pres}(a) \otimes \mathtt{abs}(a) \to 0].$$
  $$\mathtt{well\_defined}_1(V) \stackrel{\mathrm{def}}{=} \forall a \in V.\ [\mathtt{pres}(a) \oplus \mathtt{abs}(a)].$$
  $$\mathtt{well\_defined}(V) \stackrel{\mathrm{def}}{=} \mathtt{well\_defined}_0(V), \mathtt{well\_defined}_1(V).$$
- *The system*:
  $$\mathtt{vars} \stackrel{\mathrm{def}}{=} \{\mathtt{p53}, \mathtt{Mdm2}, \mathtt{DNAdam}\}.$$
  $$\mathtt{rule}(1) \stackrel{\mathrm{def}}{=} \mathtt{inhib}(\mathtt{vars}, \mathtt{DNAdam}, \mathtt{Mdm2}).\quad \mathtt{rule}(4) \stackrel{\mathrm{def}}{=} \mathtt{inhib}(\mathtt{vars}, \mathtt{Mdm2}, \mathtt{p53}).$$
  $$\mathtt{rule}(2) \stackrel{\mathrm{def}}{=} \mathtt{inhib}_s(\mathtt{vars}, \mathtt{Mdm2}, \mathtt{p53}).\quad \mathtt{rule}(5) \stackrel{\mathrm{def}}{=} \mathtt{inhib}_c(\mathtt{vars}, \mathtt{p53}, \mathtt{DNAdam}).$$
  $$\mathtt{rule}(3) \stackrel{\mathrm{def}}{=} \mathtt{active}(\mathtt{vars}, \mathtt{p53}, \mathtt{Mdm2}).\quad \mathtt{rule}(6) \stackrel{\mathrm{def}}{=} \mathtt{inhib}_s(\mathtt{vars}, \mathtt{DNAdam}, \mathtt{Mdm2}).$$
  $$\mathtt{system} \stackrel{\mathrm{def}}{=} \mathtt{vars}, \mathtt{rule}(1), \mathtt{rule}(2), \mathtt{rule}(3),$$
  $$\mathtt{rule}(4), \mathtt{rule}(5), \mathtt{rule}(6), \mathtt{well\_defined}(\mathtt{vars}).$$
- *Initial state*:
  $$\mathtt{initial\_state} \stackrel{\mathrm{def}}{=} \mathtt{abs}(\mathtt{p53}) \otimes \mathtt{pres}(\mathtt{Mdm2}),\quad \mathtt{initial\_state} \text{ at } 0.$$

Fig. 2. Representation of the System in HyLL

4.3   Proofs

Although linear logic is well suited to describing transition systems, as we do here in the area of biology, this logic can sometimes be too precise in its resource management for our needs. To solve this constraint, we sometimes make precise that we do not care about the value of some variables. We define a dont_care predicate for this purpose:

$$\mathtt{dont\_care}(x) \stackrel{\mathrm{def}}{=} \mathtt{pres}(x) \oplus \mathtt{abs}(x) \qquad \mathtt{dont\_care}(V) \stackrel{\mathrm{def}}{=} \otimes_{x \in V} \mathtt{dont\_care}(x).$$

This predicate is used in the statement of two of the four properties we present in this paper (Properties 1 and 4).

Additionally, in some proofs, when we do not know the value of some variables of the system, we sometimes need to perform a case analysis on their two possible values, using the well_defined$_1$ predicate introduced for this purpose.

Finally let us give two definitions in order to further shorten propositions and proofs ($state_0$ is a state equivalent to the initial state):

$$state_0 \overset{\text{def}}{=} \text{abs}(\text{p53}) \otimes \text{pres}(\text{Mdm2}) \qquad state_1 \overset{\text{def}}{=} \text{pres}(\text{p53}) \otimes \text{abs}(\text{Mdm2}).$$

Property 1. As long as there is DNA damage, the above system can oscillate (with a short period) from $state_0$ to $state_1$ and back again. We outline the proof informally for this property only. Others are omitted. We refer the reader to the electronic appendix (for both the sequent proofs and their formalization).

From $state_0$ and $\text{pres}(\text{DNAdam})$ we get $\text{abs}(\text{p53})$, $\text{abs}(\text{Mdm2})$, and $\text{pres}(\text{DNAdam})$ by rule 1. Then $\text{pres}(\text{p53})$, $\text{abs}(\text{Mdm2})$, and $\text{pres}(\text{DNAdam})$ ($state_1$) by rule 2. Then $\text{pres}(\text{p53})$, $\text{pres}(\text{Mdm2})$, and $\text{pres}(\text{DNAdam})$ by rule 3, and finally $\text{abs}(\text{p53})$, $\text{pres}(\text{Mdm2})$, and $\text{pres}(\text{DNAdam})$ ($state_0$) by rule 4.

We define (and prove) our property in the two possible ways discussed earlier (Sect. 4.2), roughly corresponding to Definitions 7 and 9, respectively. The difference here is that our initial state (the one from which the oscillation starts) includes the presence of DNA damage.

Proposition (Property 1, Version 1). For any world w, there exists two worlds u and v such that both u and v are less than 3 and the following holds:
$\dagger\, system @ 0 ;\, state_0 \otimes \text{pres}(\text{DNAdam}) @ w$
$\vdash \delta_u[(state_1 \otimes \text{dont\_care}(\text{DNAdam}))\, \&$
$(\delta_v(state_0 \otimes \text{dont\_care}(\text{DNAdam})))] @ w$

Alternatively, our property can be defined:

Proposition (Property 1, Version 2). For any world w, there exists two worlds u and v such that both u and v are less than 3 and the following holds:
$\dagger\, system @ 0 ;\, state_0 \otimes \text{pres}(\text{DNAdam}) @ w$
$\vdash state_1 \otimes \text{dont\_care}(\text{DNAdam}) @ w.u$ and
$\dagger\, system @ 0 ;\, state_1 @ w.u \vdash state_0 @ w.u.v$

There are no **dont_care**'s needed in the conclusion of the second sequent because only rules 3 and 4 are used, which don't involve DNAdam.

Property 2. DNA damage can be quickly recovered. This property can be stated directly as follows.

Proposition (Property 2). For any world w, there exists a world u such that u is less than 5 and the following holds:
$\dagger\, system @ 0;\, state_0 \otimes \text{pres}(\text{DNAdam}) @ w \vdash state_0 \otimes \text{abs}(\text{DNAdam}) @ w.u$

Induction/Case Analysis. Most of interesting proofs require case analysis or induction; this is the case for Properties 3 and 4 below. More precisely, we need here *case analysis on the set of fireable rules*. We implement this by a case analysis on the interval [1..6] of our six rules, together with a new predicate `fireable` defining the necessary conditions for each rule to fire. We shall also need the negation of this predicate: `not_fireable`. We give here the definitions of these predicates for the first rule of our system, for both (strong and general) styles of the rules used in the present paper (the complete definitions can be found in Appendix B):

$$\texttt{fireable}_{\texttt{s}}(1) \stackrel{\text{def}}{=} \texttt{pres(DNAdam)} \otimes \texttt{pres(Mdm2)} \otimes \texttt{dont\_care(p53)}$$

$$\texttt{not\_fireable}_{\texttt{s}}(1) \stackrel{\text{def}}{=}$$
$$((\texttt{abs(DNAdam)} \otimes \texttt{pres(Mdm2)}) \oplus (\texttt{pres(DNAdam)} \otimes \texttt{abs(Mdm2)}) \oplus$$
$$(\texttt{abs(DNAdam)} \otimes \texttt{abs(Mdm2)})) \otimes \texttt{dont\_care(p53)}$$

$$\texttt{fireable}(1) \stackrel{\text{def}}{=}$$
$$(\texttt{pres(DNAdam)} \oplus (\texttt{pres(DNAdam)} \otimes \texttt{pres(Mdm2)}) \oplus$$
$$(\texttt{pres(DNAdam)} \otimes \texttt{abs(Mdm2)})) \otimes \texttt{dont\_care(p53)}$$

$$\texttt{not\_fireable}(1) \stackrel{\text{def}}{=} \texttt{abs(DNAdam)} \otimes \texttt{dont\_care(\{Mdm2, p53\})}$$

An (informal) formula like "for any *fireable rule r, P*" will be written as "for any rule r in the interval [1..6], the following holds: $(\texttt{fireable}(r) \ \& \ P) \oplus \texttt{not\_fireable}(r)$". Note that both definitions of `fireable` and `not_fireable` could be generated from the definitions of the biological rules, as the rules we consider in the present paper always have the same shape.

Property 3. If there is no DNA damage, the system remains in the initial state. A first attempt at formalizing this property might be:

For any world w, the following holds: $\dagger system @ 0, \texttt{abs(DNAdam)} @ 0 \vdash state_0 \otimes \texttt{abs(DNAdam)} @ w$.

However, the above statement does not model our property. We want to prove that if $\texttt{abs(DNAdam)} @ 0$ then $state_0 \otimes \texttt{abs(DNAdam)} @ w$ holds, for all worlds w, *no matter which rule is fired* to get to w. Thus our property requires a *case analysis* on the rules of the biological system.

Proposition (Property 3). *Let $\mathcal{P}$ denote the formula $state_0 \otimes \texttt{abs(DNAdam)}$. For any world w, the following holds: $\dagger system @ 0, \mathcal{P} @ 0 \vdash \mathcal{P} \ at \ 0 @ w$; and for any world w, for any rule r in the interval [1..6], the following holds:*

$$\dagger system @ 0 \vdash \mathcal{P} \rightarrow (\texttt{fireable}(r) \ \& \ \delta_1 \mathcal{P}) \oplus \texttt{not\_fireable}(r) @ w$$

The proof of the second statement proceeds by case analysis on the rules (r) of the biological system. There are only two fireable rules: $\texttt{rule}(4)$ and $\texttt{rule}(6)$.

Property 4. There is no path with two consecutive states where p53 and Mdm2 are both present or both absent. In other words: from any state where p53 and

Mdm2 are both present or both absent, we can only go to a state where either p53 is present and Mdm2 is absent or p53 is absent and Mdm2 is present.

This requires a stronger (natural) hypothesis: we need the property that each rule modifies at least one entity in the system. In order to achieve this, we shall use the strong style of definitions for our inhibition and activation rules discussed earlier (Sect. 4.2). For example, the activation rule will be defined as follows:

$$s\_active(V, a, b) \stackrel{\text{def}}{=} pres(a) \otimes abs(b) \rightarrow$$
$$\delta_1(pres(a) \otimes pres(b)) \otimes \downarrow u.\ unchanged(V \setminus \{a, b\}, u)).$$

The complete set of strong rules can be found in Appendix B.

Let $\mathcal{L}$ and $\mathcal{R}$ denote the following two formulas:

$$\mathcal{L} := (\text{pres}(p53) \otimes \text{pres}(Mdm2)) \oplus (\text{abs}(p53) \otimes \text{abs}(Mdm2))$$
$$\mathcal{R} := ((\text{pres}(p53) \otimes \text{abs}(Mdm2)) \oplus$$
$$(\text{abs}(p53) \otimes \text{pres}(Mdm2))) \otimes \text{dont\_care}(DNAdam)$$

We want to prove that from state $\mathcal{L}$ we can only go to state $\mathcal{R}$, *no matter which rule is fired*. Here again, we need *case analysis on the set of fireable rules*:

Proposition (Property 4). *For any world w, for any rule r in the interval* $[1..6]$, *the following holds:*

$$\dagger\, system\ @\ 0;\ .\vdash \mathcal{L} \rightarrow (s\_fireable(r)\ \&\ \delta_1\,\mathcal{R}) \oplus s\_not\_fireable(r)\ @\ w$$

Property 4 could be written as the CTL formula $\mathsf{AG}(\mathcal{L} \rightarrow \mathsf{AX}\mathcal{R})$ (see Sect. 6 for the encoding of such a formula in HyLL). Nevertheless, to simplify the proof we can observe that the argument property of AG contains an implication and thus all the possible states verifying the left hand side of the implication should be taken into account in the proof. As a matter of fact, at each step we do not make assumptions on the state where $\mathcal{L}$ holds. The system satisfies Property 4 if all its states satisfy $\mathcal{L} \rightarrow \mathsf{AX}\mathcal{R}$. Since in our transition system all the states are reachable from the initial states, this corresponds to requiring the satisfaction of Property 4 at (that is, from) the initial states. In case we want to test such a property at a state S_i that is not connected to all the other states, we only need to prove the property in the subtree of the transition system rooted in S_i. In HyLL, we should prove the following theorem: *if* $reachable(S, S_i)$ *then* $S \vdash \mathcal{L} \rightarrow \forall r \in R.\ \delta_1 \mathcal{R}$, where $reachable(S, S_i) \stackrel{\text{def}}{=} \exists u.\ S_i \rightarrow \delta_u S$.

5   Formal Proofs

Our approach is to fully formalize proofs in Coq, using the λProlog prover to help with partial automation of the proofs. The λProlog prover is a tactic-style interactive theorem prover implemented in the manner described in [16], also given as an example in Sect. 9.4 of [27]. The code described there can be viewed as a logical framework, implementing a tactic-style architecture. We use this code directly, and instantiate the framework with tactics implementing the basic inference rules of HyLL.

We use Coq for two reasons. First, we can build libraries in Coq that allow us to reason at two levels. We can prove meta-level properties of HyLL (for example, we have formalized Theorem 2), and we can reason at the object-level, which in this case means that we can prove HyLL sequents directly. To do so, we adopt the two-level style of reasoning used in Hybrid [17] where the logic we want to reason in and about is called the *specification logic* and is implemented as an inductive predicate in Coq. The inductive predicate in this case defines HyLL sequents, and its definition is a fairly direct modification of the ordered linear logic in [17]. Second, once a proof is complete, Coq provides a *proof certificate*. In particular, Coq implements the calculus of inductive contstructions (CIC), where a property is stated as a type in CIC and a proof is a λ-term inhabiting that type. This λ-term serves as a certificate, which can be stored and checked independently.

It is, of course, possible to prove the properties in Sect. 4 only using Coq, but in general, proofs in Coq of HyLL sequents quickly become cumbersome because of the amount of detail required to apply each inference rule of HyLL. The λProlog prover is used to automatically infer much of this detail. For example, because we implement HyLL directly as an inductive predicate in Coq, in order to apply the $\otimes L$ rule (see Fig. 1), Coq's `apply` tactic requires arguments to be given explicitly for the instantiation of formulas A and B, world u, and multisets Δ and $\Delta, A @ u, B @ u$. (Using the more flexible `eapply` does not help with the proofs considered here.) As a result, the Coq proofs are verbose and often contain redundant information. In λProlog on the other hand, our primitive tactics for applying HyLL inference rules use unification to infer these arguments. We could instead program a more automated version of the `apply` tactic in Coq, tailored to applying HyLL rules using Coq's Ltac facility, but this task would likely be more complex and adhoc, since unification is not one of the primitive operations of Coq.

It is also straightforward to represent HyLL proof terms in λProlog, and implement the construction of these proof terms as part of the implementation of the primitive tactics. In our λProlog prover, we construct such proof terms and then translate them to strings representing Coq proof script. This translation is also implemented in λProlog. These automatically generated proof scripts are imported into Coq, and then after some fine-tuning of the script, we obtain a proof certificate for the entire proof.

Note that we do not use λProlog to fully automate proofs; the λProlog prover is also interactive. The user indicates what HyLL rule to apply at each step, possibly with some other basic information such as the position in Δ of the formula to which the rule is to be applied. The arguments needed to apply the rule are then inferred automatically. We could build more automation into the prover, possibly in the style of the linear logic programming language presented in [21] for general automation, with the addition of heuristics tailored to our application. This is left for future work.

The reader is referred to the electronic appendix for the encoding of our biological system in Coq. Here, for illustration purposes, we simply state and discuss Property 3 exactly as it appears in Coq.

```
Theorem Property3 : forall w:world,
  seq Gamma ((PP @ 0)::nil) ((PP at 0) @ w) /\
  forall (n:nat) (A B:oo_), fireable n A -> not_fireable n B ->
  seq Gamma nil ((PP ->> ((A &a step PP) +o B)) @ w).
```

In general, sequents have the form (seq Gamma Delta (A @ W)), where A represents a HyLL formula (of type oo_ in the Coq encoding) and W is a world, which are encoded using Coq's built-in type for natural numbers. Gamma is a list of HyLL formulas (using the built-in lists of Coq) and Delta is a multiset of elements of type oo_, where we build our own custom multiset library. In the above theorem, Gamma is the Coq encoding of († *system* @ 0). In the first sequent in the statement of the theorem, Delta contains only (PP @ 0) where PP is the Coq encoding of ($state_0$ $\otimes$ abs(DNAdam)), and in the second sequent Delta is empty. The symbols and constants @, at, ->>, &a, step, and +o represent the HyLL operators @, at, $\rightarrow$, &, δ_1, and $\oplus$, respectively. The predicates fireable and not_fireable are defined inductively in Coq, where the first argument is a natural number specifying the rule number.

The Coq proof of the second conjunct proceeds by case analysis on n, followed by inversion on (fireable n A) and (not_fireable n B), which provides instantiations for A and B (the conditions that express whether the rule is fireable or not). The resulting 6 subgoals are sent to the λProlog prover, whose output is imported back into Coq as described above.

6   Comparison with Model Checking

While temporal logics such as LTL, CTL, or CTL* have been very successful in practice with efficient model checking tools, the proof theory of these logics is very complex. In contrast, HyLL has a very traditional proof theoretic pedigree: it is presented in the sequent calculus and enjoys cut-elimination and focusing [8]. A further advantage of our approach with respect to model checking is that it provides an unified framework to encode both transition rules and (both statements and proofs of) temporal properties.

Let us examine both approaches in more details.

6.1   Temporal Operators

We propose the following encoding of temporal logic operators in HyLL[$\mathcal{T}$], where $\mathcal{T} = \langle I\!N, +, 0 \rangle$, representing instants of time. While this domain does not have any branching structure like CTL, it is expressive enough for many common idioms because of the branching structure of derivations involving $\oplus$.

State quantifiers can be easily mapped. There is a clear correspondence between F (resp. G) and $\Diamond$ (resp. $\Box$), see Definition 6. The encodings of X and U are the following ones: $\mathsf{X}P \Leftrightarrow \delta_1 P$ and $P_1 \mathsf{U} P_2 \Leftrightarrow\downarrow u.\ \exists v.\ P_2$ at $u.v \otimes \forall w <$ $v.\ P_1$ at $u.w$. where P, P_1, and P_2 are some propositions (not necessarily atomic ones).[1]

As for path quantifiers, the question is more subtle. The idea is that E corresponds to the existence of a proof, while for A it is necessary to look at a proof considering all the possible rules to be applied at each step (at each step of the proof, the chosen rule should not influence the property satisfaction).

---

[1] Observe that a proposition characterizes a set of states.

We came to the conclusion that the encoding of A in HyLL depends on the state quantifier following it. Let R be the set of rules of our transition system. The mapping we propose is the following one:

- $\mathsf{AX}P$. In HyLL, we write $\forall r \in R \, \delta_1 P$. More precisely, the encoding contemplating fireable rules is $\forall r \in R \, (\mathtt{fireable}(r) \, \& \, \delta_1 P) \oplus \mathtt{not\_fireable}(r)$ (see Sect. 4.3). For the sake of simplicity, in the following we omit such details concerning fireable rules.
- $\mathsf{AG}P$. It is equivalent to $P \wedge \mathsf{AG}(P \to \mathsf{AX}(P))$. In HyLL, we write $P \otimes \forall n (P \, at \, n) \to \forall r \in R(P \, at \, n+1)$.
- $\mathsf{AF}P$. It is equivalent to $P \vee \mathsf{AX}(\mathsf{AF}P)$. If we have a bound k on the number of steps needed to satisfy the property, we can expand this formula by obtaining: $P \vee \mathsf{AX}(P \vee \mathsf{AX}(\dots \mathsf{AX}P))$, with k nested occurrences of AX. In HyLL, we write $P \oplus \forall r \in R(\delta_1 P \oplus (\forall r \in R(\dots \delta_k P)))$. Notice that another alternative is to express the F operator by using U ($\mathsf{F}P \Leftrightarrow \mathsf{true} \, \mathsf{U}P$).
- $\mathsf{A}(P_1 \mathsf{U} P_2)$. It is equivalent to $P_2 \vee (P_1 \wedge \mathsf{AX}(P_1 \mathsf{U} P_2))$. If we have a bound k on the number of steps needed to satisfy the property, we can expand this formula by obtaining: $P_2 \vee (P_1 \wedge \mathsf{AX}(P_2 \vee (P_1 \wedge \mathsf{AX}(\dots \mathsf{AX}P_2))))$, with k nested occurrences of AX. In HyLL, we write $P_2 \oplus (P_1 \otimes \forall r \in R(\delta_1 P_2 \oplus (\delta_1 P_1 \otimes \forall r \in R(\dots \delta_k P_2))))$.

In addition to the future connectives, the domain $\mathcal{T}$ also admits past connectives if we add saturating subtraction (i. e., $a - b = 0$ if $b \geq a$) to the language of worlds. We can then define the duals H (historically) and O (once) of G and F as:

$$\mathsf{H} \, P \stackrel{\mathrm{def}}{=} \downarrow u.\forall w.(P \, \mathsf{at} \, u - w) \qquad \mathsf{O} \, P \stackrel{\mathrm{def}}{=} \downarrow u.\exists w.(P \, \mathsf{at} \, u - w)$$

6.2   Model Checking

A strength of our approach with respect to model checking is that, when we prove an existential property using certain rules of a model, we have the guarantee that *all* the models containing such rules satisfy the property. This is important because in biology we often deal with incomplete information. It is also worth noting that in model checking, all objects are finite: both the number of states, and the number of transitions in the state graph. In HyLL, objects can potentially be infinite; in particular, we can have an infinite number of states. Let us point out further advantages of our approach with respect to model checking. First of all, when proving a given property we do not need to blindly try all possible rules at each step but we can guide the proof (see Sect. 7). Observe that a successful proof of a given property can be exploited to prove similar properties. Furthermore, suppose we are able to prove a property of the system which is not desirable. In this case the proof we get can help us in understanding what should be modified in the system so that the property is not satisfied. More precisely, we can look for the rules to be removed/modified among those that have been used in the proof. In model checking, when a property turns out to be true, the reason is not investigated. Finally, in [34], temporal logic is extended to allow the expression of properties such as "P is true at every even state of an infinite path." A decision procedure for this extended logic is also defined, but to the best of our knowledge, there is no model checking tool for it. In HyLL, if we add equality on worlds, we can write $\forall n = 2k. \ P \ at \ n$.

Note that in some of the temporal properties we test, there is a bound on the number of time units, and thus on the length of the proof. In this particular case, there is a strong analogy with "bounded model checking" [3].

A drawback of theorem proving with respect to model checking is that this method can be time consuming and needs an expert. Recent advances in both proof theory and systems however provide us with at least partial, and sometimes complete, automation of the proofs.

7   Conclusion and Future Work

In this paper we argued that the HyLL logic can be successfully exploited for formally verifying Boolean biological systems. This work is a first experiment along this new line of research (although we already provide fully mechanized proofs). We focussed on a simple regulatory network but our framework could be adopted to model several other kinds of biological networks (e.g., neuronal, predator-prey, or ecological networks).

A natural extension of this work consists of applying our methodology to multivalued, continuous, and stochastic biological models. As far as the first case is concerned, the extension is straightforward, we just need to replace present/absent predicates by predicates indicating the discrete values of variables. The latter two cases are more involved. We could try to use the domain of world $\mathcal{W} = \langle \Re^+, +, 0 \rangle$, use predicates to represent variable concentrations and express the evolution of each variable in terms of functions involving kinetic expressions such as the mass action law, Hill, or Michaelis-Menten kinetics. In [8], several alternatives for the worlds in the probabilistic case are also discussed. In any case, the challenge would consist of being able to perform symbolic calculation as much as possible without evaluating functions.

Proofs of properties such as 1 and 2 require finding a path through the system, which here means specifying a series of rules that can be applied in a particular order. At each step, there may be a choice between several potential fireable rules. In the interactive proofs, such choices were made by hand. Our future work includes building automated procedures (e.g., Coq tactics) to guide the proof. We would also like to extend our model to include axioms for events such as those considered in Biocham, which make it possible to change the value of some variables under certain special conditions. Such events often correspond to external inputs and have priority over the ordinary rules of a model.

Our aim is to find the logical essence of biochemical reactions. What we envision for the domain of "biological computation" is a resource-aware stochastic or probabilistic λ-calculus that has HyLL propositions as (behavioral) types.

References

1. Alur, R., Belta, C., Ivančić, F., Kumar, V., Mintz, M., Pappas, G.J., Rubin, H., Schug, J.: Hybrid modeling and simulation of biomolecular networks. In: Di Benedetto, M.D., Sangiovanni-Vincentelli, A.L. (eds.) HSCC 2001. LNCS, vol. 2034, pp. 19–32. Springer, Heidelberg (2001)
2. Bertot, Y., Castéran, P.: Interactive Theorem Proving and Program Development. In: Coq'Art: The Calculus of Inductive Constructions. Springer (2004)
3. Biere, A., Cimatti, A., Clarke, E.M., Strichman, O., Zhu, Y.: Bounded model checking. In: Zelkowitz, M. (ed.) Highly Dependable Software. Advances in Computers, vol. 58, pp. 117–148. Elsevier (2003)
4. Bozzano, M.: A Logic-Based Approach to Model Checking of Parameterized and Infinite-State Systems. Ph.D. thesis, DISI, Università di Genova (2002)
5. Campagna, D., Piazza, C.: Hybrid automata in systems biology: How far can we go? Electronic Notes in Theoretical Computer Science 229(1), 93–108 (2009)
6. Cervesato, I., Pfenning, F., Walker, D., Watkins, K.: A concurrent logical framework II: Examples and applications. Tech. Rep. CMU-CS-02-102, Carnegie Mellon University (2003)
7. Chabrier-Rivier, N., Chiaverini, M., Danos, V., Fages, F., Schächter, V.: Modeling and querying biochemical interaction networks. Theoretical Computer Science 325(1), 25–44 (2004)
8. Chaudhuri, K., Despeyroux, J.: A hybrid linear logic for constrained transition systems with applications to molecular biology. Tech. Rep. inria-00402942, INRIA-HAL (October 2013)
9. Ciliberto, A., Novák, B., Tyson, J.J.: Steady states and oscillations in the p53/Mdm2 network. Cell Cycle 4(3), 488–493 (2005)
10. Cimatti, A., Clarke, E., Giunchiglia, F., Roveri, M.: NUSMV: A new symbolic model verifier. In: Halbwachs, N., Peled, D.A. (eds.) CAV 1999. LNCS, vol. 1633, pp. 495–499. Springer, Heidelberg (1999)
11. Clarke, J.E.M., Grumberg, O., Peled, D.A.: Model checking. MIT Press, Cambridge (1999)
12. Courcoubetis, C., Vardi, M.Y., Wolper, P., Yannakakis, M.: Memory-efficient algorithms for the verification of temporal properties. Formal Methods in System Design 1(2/3), 275–288 (1992)
13. Danos, V., Laneve, C.: Formal molecular biology. Theoretical Computer Science 325(1), 69–110 (2004)
14. Despeyroux, J., Chaudhuri, K.: A hybrid linear logic for constrained transition systems. To Appear in Types for Proofs and Programs, Post-Proceedings of TYPES 2013. LIpIcs (Leibniz International Proceedings in Informatics) (2014)
15. Fages, F., Soliman, S., Chabrier-Rivier, N.: Modelling and querying interaction networks in the biochemical abstract machine BIOCHAM. Journal of Biological Physics and Chemistry 4(2), 64–73 (2004)
16. Felty, A.: Implementing tactics and tacticals in a higher-order logic programming language. Journal of Automated Reasoning 11(1), 43–81 (1993)
17. Felty, A.P., Momigliano, A.: Hybrid: A definitional two-level approach to reasoning with higher-order abstract syntax. Journal of Automated Reasoning 48(1), 43–105 (2012)
18. Gentzen, G.: Investigations into logical deductions. In: Szabo, M.E. (ed.) The Collected Papers of Gerhard Gentzen, pp. 68–131. North-Holland Publishing Co., Amsterdam (1969)

19. Girard, J.Y.: Linear logic. Theoretical Computer Science 50, 1–102 (1987)
20. Hinton, A., Kwiatkowska, M., Norman, G., Parker, D.: PRISM: A tool for automatic verification of probabilistic systems. In: Hermanns, H., Palsberg, J. (eds.) TACAS 2006. LNCS, vol. 3920, pp. 441–444. Springer, Heidelberg (2006)
21. Hodas, J.S., Miller, D.: Logic programming in a fragment of intuitionistic linear logic. Journal of Information and Computation 110(2), 327–365 (1994)
22. Hofestädt, R., Thelen, S.: Quantitative modeling of biochemical networks. In: Silico Biology, vol. 1, pp. 39–53. IOS Press (1998)
23. Holzmann, G.J.: The Spin Model Checker: Primer and Reference Manual. Addison-Wesley Professional (2003)
24. de Jong, H., Gouzé, J.L., Hernandez, C., Page, M., Sari, T., Geiselmann, J.: Qualitative simulation of genetic regulatory networks using piecewise-linear models. Bulletin of Mathematical Biology 66(2), 301–340 (2004)
25. Maria, E.D., Fages, F., Rizk, A., Soliman, S.: Design, optimization and predictions of a coupled model of the cell cycle, circadian clock, DNA repair system, irinotecan metabolism and exposure control under temporal logic constraints. Theoretical Computer Science 412(21), 2108–2127 (2011)
26. Miller, D.: The π-calculus as a theory in linear logic: Preliminary results. In: Lamma, E., Mello, P. (eds.) ELP 1992. LNCS, vol. 660, pp. 242–265. Springer, Heidelberg (1993)
27. Miller, D., Nadathur, G.: Programming with Higher-Order Logic. Cambridge University Press (2012)
28. Phillips, A., Cardelli, L.: A correct abstract machine for the stochastic pi-calculus. In: BioConcur: Workshop on Concurrent Models in Molecular Biology. Electronic Notes in Theoretical Computer Science (2004)
29. Reddy, V.N., Mavrovouniotis, M.L., Liebman, M.N.: Petri net representations in metabolic pathways. In: First International Conference on Intelligent Systems for Molecular Biology, pp. 328–336. AAAI Press (1993)
30. Regev, A., Panina, E.M., Silverman, W., Cardelli, L., Shapiro, E.: Bioambients: An abstraction for biological compartments. Theoretical Computer Science 325(1), 141–167 (2004)
31. Regev, A., Silverman, W., Shapiro, E.Y.: Representation and simulation of biochemical processes using the π-calculus process algebra. In: Sixth Pacific Symposium on Biocomputing, pp. 459–470 (2001)
32. Thomas, R.: Boolean formalization of genetic control circuits. Journal of Theoretical Biology 42(3), 563–585 (1973)
33. Thomas, R., Thieffry, D., Kaufman, M.: Dynamical behaviour of biological regulatory networks—I. Biological Role of Feedback Loops and Practical Use of the Concept of the Loop-Characteristic State. Bulletin of Mathematical Biology 57(2), 247–276 (1995)
34. Wolper, P.: Temporal logic can be more expressive. Information and Control 56(1/2), 72–99 (1983)

Disentangling the Effects of Habitat and Protection on Coral Reef Fish Communities in Long-Established Marine Reserves

Dominique Pelletier[1], Delphine Mallet[1,2], Abigail Powel[1], and William Roman[1]

[1] IFREMER, UR Lagons Ecosystèmes & Aquaculture Durable
Nouméa, New Caledonia
`firstname.surname@ifremer.fr`
[2] Université de la Nouvelle-Calédonie
Nouméa, New Caledonia

Abstract. Marine Protected Areas (MPA) play a central role in policies for the conservation of coastal ecosystems and resources. One of the current challenges of monitoring the effectiveness of MPAs at protecting fish assemblages is distinguishing between the influence of habitat, protection effects and anthropogenic activities. We used strongly spatially-replicated data to investigate these effects within an area comprising two no-take long-established marine reserves. Data were obtained from remote underwater video collected across all habitats in the study area. First, each station was assigned to a habitat based on a multivariate typology of biotic and abiotic cover. A number of metrics including fish abundances, assemblage composition and trophic structure were then modeled as a function of habitat and protection status. We showed that protection effects may be easily detected from a number of metrics involving fished species, but for other metrics, effect detection requires that habitat is explicitly taken into account.

Keywords: Marine Protected Area, monitoring, indicators, habitat, coral reefs, fish assemblages.

1 Introduction

Marine reserves and more generally Marine Protected Areas (MPA) have become a central management tool for the conservation and restoration of marine ecosystems and resources. Evaluation of the success or failure of an MPA to meet these objectives is critical in order to provide managers with the elements necessary to reinforce MPA acceptance or to adapt MPA design for a better achievement of objectives. In the Southwest lagoon of New Caledonia, no-take marine reserves were established around five islets in 1989. These islands are located near to the city of Noumea and thus experience high levels of recreational use. Wantiez et al. (1997) documented the effects of protection on reef fish density, diversity and biomass within these reserves [2]. In this work, we use strongly spatially-replicated data across a number of reef habitats to investigate the effects of

F. Fages and C. Piazza (Eds.): FMMB 2014, LNBI 8738, pp. 156–158, 2014.
© Springer International Publishing Switzerland 2014

habitat and protection on a number of metrics related to fish assemblages in a highly diversified coral reef ecosystem. We aim to disentangle these factors and examine their effects on a number of univariate and multivariate metrics within an area where protection has been in place for more than thirty years.

2 Material and Methods

The study area was located in the Southwest Lagoon of New Caledonia, South Pacific (22°22.5′S, 166°14′E). This area includes two protected areas where all fishing has been prohibited since 1989, and three nearby unprotected reefs, as well as in the lagoon between the two protected islets. Remote Underwater Video observations were conducted in the various habitats around these islets and reefs, including coral reef areas, seagrass beds and soft-bottom areas using the STAVIRO observation system described by [1]. Videos were analysed to identify and count fish and marine turtles to the most precise taxonomic level possible. For each species, resulting counts were used to compute abundance density per species at each station. A larger set of data collected across the New Caledonian lagoon (1450 observations in total) was analysed to build a typology of stations based on biotic and abiotic cover, depth, topography and complexity based on Hierarchical Ascending Clustering. Fish presence and density were analysed to investigate and explain spatial patterns considering habitat, reef type, protection status as explanatory factors. Finally, the effects of habitat and protection status on the entire fish assemblage were analysed using permutational multivariate analysis of variance (PERMANOVA). Analyses were based on resemblance matrices calculated using Bray-Curtis similarity coefficients. Differences in fish assemblages between stations were tested using two-way PERMANOVA with factors protection status and habitat.

3 Results

No significant differences between protected and unprotected areas could be detected from overall fish density and species richness. In contrast, the density of spearfished species and the density of herbivores showed a significant protection effect. Detection of protection effects was improved when habitat was considered in the models. Most metrics then displayed higher values in the protected areas than outside, in particular, the densities of species caught by speargun, by net and by line, and the densities of herbivores, piscivores and carnivores. The reverse was true for plankton feeders.

In a second step, we focused on three important fished species. Blue spine surgeonfish (Naso unicornis) density was significantly higher in reef habitats. Spangled emperor (Lethrinus nebulosus) density was higher in sea grass habitats in protected areas, although not significantly ($p < 0.12$). Coral trout (Plectropomus leopardus), a major target species for spearfishers was encountered in the live coral and debris habitat. In both habitats, P. leopardus densities were higher in the protected area, although not significantly, as $p < 0.055$ in the living coral habitat, and $p < 0.16$ in the debris habitat.

At the fish assemblage level, species composition and the trophic structure of fish assemblages were significantly affected by protection status and habitat. The effect of protection status on assemblage structure differed according to habitat. In the debris habitat, we found significant differences in species assemblages, family composition and trophic structure in debris habitats inside and outside reserves. In coral habitats, we found significant differences in species assemblages and family composition inside and outside reserves. No significant effects of protection were observed on any of the metrics in the soft bottom habitats.

4   Conclusion

One of current challenges of monitoring MPA effects is the need to take into account the effects of habitat on reef fish assemblages. By using a remote underwater video method that allows a high degree of spatial replication and sampling design stratified according to geomorphology of reef (reef type) and protection status, we were able to survey fish assemblages and habitats at a large number of stations. Habitat data collected concomitantly with fish data were used to characterize station habitat at a more precise scale than reef type only. The effects of habitat and protection on fish assemblages could then be investigated across a variety of habitats. Significant effects of protection and/or habitat were detected for several metrics, including overall species densities or species richness, the densities of commercially important species and at the fish assemblage level. Our results support the findings that protection effects are more apparent on fish densities when habitat is taken into account but also highlight the importance of sampling reef associated habitats.

References

1. Pelletier, D., Leleu, K., Mallet, D., Herve, G., Mou Tham, G., et al.: High-Definition Rotating Video Enables Fast Spatial Survey of Marine Underwater Macrofauna and Habitats. PLoS ONE 7(2), e30536 (2012)
2. Wantiez, L., Thollot, P., Kulbicki, M.: Effects of marine reserves on coral reef fish communities from five islands in New Caledonia. Coral Reefs 16, 215–224

The Challenges of Developing Spatially Explicit Network Models for the Management of Disease Vectors in Ecological Systems

Brendan Trewin[1,2], Hazel Parry[1], Myron Zalucki[2],
David Westcott[1], and Nancy Schellhorn[1]

[1] CSIRO, Brisbane, Australia
[2] University of Queensland, Brisbane, Australia

Challenges of modelling vector-borne disease systems result from complexities and uncertainities inherent in the vector's behavioural ecology and its interactions in a landscape context. Network models provide a number of approaches and measures to quantify spatially-explicit systems that are consistent with the ecological process of vector dispersal, with implications for disease transmission and spread [1,2]. Here we discuss two spatially explicit vector systems as network models; (1) the movement of the invasive mosquito *Aedes aegypti*, which vectors a number of diseases including dengue fever, through rainwater tanks in a major urban area, (2) the movement of bats (flying-foxes), which vector Hendra virus, through urban and rural landscapes [3]. We contrast the design and applicability of these networks, comparing features and challenges inherent in modelling these systems, and discuss the use of network models as disease vector management tools with implications for disease spread.

In an ecological context, nodes often represent metapopulations and compartmentalize important demographic characteristics such as growth rate, disease transmission rate and spatial location within landscapes. In our mosquito model, rainwater tanks are nodes that are fixed in both space and time, with accurate location data available from government rebate schemes. Depending on whether nodes are exposed to the environment (non-compliant) or not, tanks are nodes that may act as sources or sinks for mosquito vectors respectively. Characteristics that govern population growth within each source node are simple to collect and model as there is a vast literature on simulating population growth within containers [4]. Within the bat-Hendra model nodes are likely to be bat camps (roosts), containing populations of vectors. The highly seasonal nature of camps and their susceptibility to variations in environment and climate result in uncertainties in spatial location of the camps. Bats have high dispersal abilities with complex movement and social behaviours. This leads to large fluctuations in the formation and removal of nodes through space and time. Foraging sites could be additional nodes within this system, but are difficult to model explicitly due to their inherent stochasticity and have so far been ignored. Important simplifying assumptions are made in characterizing bat camps as nodes in a network model compared to rainwater tanks, as the tanks better reflect our compartmentalized concept of 'nodes' in a network. These assumptions introduce uncertainty into any conclusions that are drawn about the bat disease vector system, but this uncertainty is not made explicit.

F. Fages and C. Piazza (Eds.): FMMB 2014, LNBI 8738, pp. 159–161, 2014.
© Springer International Publishing Switzerland 2014

Within a network model, edges are characterized as the flow of information between nodes and it is important that these connections reflect the scale at which nodes interact. In an ecological context this is represented by a dispersal mechanism, typically either a binary variable or a continuous function that decays with distance. In our mosquito model, edges represent the movement of mosquito vectors between rainwater tanks and are characterized through a dispersal kernel algorithm that decays with increasing distance. In this way, we connect nodes by considering topology and a distance threshold based on known dispersal rates and population size. Initially, tanks may become breeding sites and put neighbouring tanks at risk of hosting vectors when deemed non-compliant. Edges within the bat-Hendra model are similarly characterized by a distance weighted 'connectivity' function between camp nodes. In reality, the magnitude and direction of bat movement between camps and foraging sites could be represented by a large number of links and are not necessary driven solely by distance. However, there are difficulties in collecting movement data with telemetry equipment and in accessing and monitoring bat camps, foraging sites and the seasonal nature of bat movement between sites. Attempting to simulate a bat vector system with a network model is therefore fraught with difficulties in obtaining accurate data to quantify the scales at which vector movement and interactions occur.

The importance of node connectivity can be explored in an ecological context by calculating measures of diffusion and node centrality within the system. The connectivity within the mosquito system allows for a higher probability of colonization if populations are large, but is also constrained by the limited dispersal ability of the species. Measures of the number of node links and node influence on a network can indicate the risk of individual nodes as disproportionate sources of infection. The ability to effectively identify and target high risk nodes or collections of nodes is considered an important goal for vector reduction (therefore reducing disease risk) by mosquito control authorities. Connectivity within the model bat-Hendra system is based on a distance weighted probability of infected individuals moving between camps. The result of the model was that the highly connected urban camps are predicted to experience small, high frequency epidemics, occasionally sprouting travelling waves of infection linearly through rural populations [3]. How well this represents disease spread through this system is very uncertain due to the bat's high dispersal ability and complex nature of movement. Until a better understanding of the dynamics within the bat system is developed, this network model is best used alongside empirical studies as a hypothesis generating tool [5].

The large contrasts between these two disease vector networks relate to how well each model represents reality. Thus when considering whether to adopt a network modelling approach, one should consider how well studied the disease vector's behavioural ecology is, as well as its interactions with the disease and the environment in space and time. Ideally, when developing network models for biosecurity or public health authorities as management tools, suitable ecological systems to simulate are those that minimize temporal/spatial stochasticity in network design, have access to accurate spatial data and give realistic insights into vector dispersal. If assumptions generate open-ended hypotheses, there may be more value to authorities to reframe questions or consider other modelling approaches.

References

1. Brooks, C.P., Antonovics, J., Keitt, T.H.: Spatial and Temporal Heterogeneity Explain Disease Dynamics in a Spatially Explicit Network Model. Amer. Nat. 172, 149–159 (2008)
2. Ferrari, J.R., Preisser, E.L., Fitzpatrick, M.C.: Modeling the spread of invasive species using dynamic network models. Biol. Invasions. 16, 949–960 (2014)
3. Plowright, R.K., Foley, P., Field, H.E., Dobson, A.P., Foley, J.E., Eby, P., Daszak, P.: Urban habituation, ecological connectivity and epidemic dampening: the emergence of Hendra virus from flying foxes (Pteropus spp. Proc. R. Soc. B. 278, 3703–3712 (2011)
4. Focks, D.A., Haile, D.G., Daniels, E., Mount, G.A.: Dynamic Life Table Model for *Aedes aegypti* (Diptera: Culicidae): Analysis of the Literature and Model Development. J. Med. Entomol. 30, 1003–1017 (1993)
5. Restif, O., Hayman, D.T.S., Pulliam, J.R.C., Plowright, R.K., George, D.B., Luis, A.D., Cunningham, A.A., Bowen, R.A., Fooks, A.R., O'Shea, T.J., Wood, J.L.N., Webb, C.T.: Model-guided fieldwork: practical guidelines for multidisciplinary research on wildlife ecological and epidemiological dynamics. Ecol. Lett. 15, 1083–1094 (2012)

Evaluating Management Scenarios for Fished Resources of the New Caledonian Lagoon Using a Spatially-Explicit Model

Bastien Preuss[1,2], Dominique Pelletier[2], and Laurent Wantiez[1]

[1] EA 4243, LIVE, Université de la Nouvelle-Calédonie
Nouméa, New Caledonia
bastien.preuss@gmail.com, laurent.wantiez@univ-nc.nc
[2] IFREMER, UR LEAD
BP 2059, 98846 Nouméa Cedex, New Caledonia
dominique.pelletier@ifremer.fr

Abstract. In the New Caledonian lagoon, fish populations live in a highly-fragmented habitat and seascape, and many are exploited by commercial, recreational and subsistence fishers. Although fishing has been increasing over decades, fisheries sustainability has not been assessed. We used the ISIS-Fish tool to build a spatially-explicit dynamic model for two major fish resources: spangled emperor and coral trout. The model was constructed and parameterized from many existing data habitat, fish populations, and fishing pressures. After calibration, it was used to simulate the outcomes of several management scenarios including MPA, size limitation, and increase in commercial fisher number. Results showed that depending on the species, an MPA could result in a significant resource increase within its boundaries. Size limitation highly reduced line fishing catches, but its benefits depended on the survival rate of released fish. Increasing the number of commercial fishers significantly affected spangled emperor but not coral trout.

Keywords: fish population dynamics, spatially-explicit model, management scenario assessment, spangled emperor, coral trout, New Caledonian lagoon.

1 Introduction

Coastal areas are subject to increasing demographic pressure, and they host activities other than fisheries. Integrating principles of sustainable development to limit loss of environmental resources is a challenge for the forthcoming years, as e.g. stated in the Millenium Goals (http://www.un.org/milleniumgoals). To reach this goal, it is necessary to implement management options that ensure fisheries sustainability. Marine Protected Areas (MPA) constitute a key policy for the management of coastal fisheries and ecosystems, because zoning of uses is often indispensable to achieve apparently conflicting goals such as biodiversity conservation and sustainable development of economic activities, including

F. Fages and C. Piazza (Eds.): FMMB 2014, LNBI 8738, pp. 162–164, 2014.

fisheries. Other management options as gear restriction or limitation of size of professional fisheries are also frequently used by decision makers.

In New Caledonia, two thirds of the human population live in or close to Noumea city. Due to fishing pressure in the surrounding lagoon, conservation measures such as no-take marine reserves, have long been established to prevent excessive fishing pressure. Yet, the sustainability of the current exploitation level has not been assessed, particularly under a growing demographic pressure.

The consequences of additional and alternative management options can be investigated using e.g. simulation models of fisheries dynamics. Spatially explicit population dynamic models are particularly relevant in this context because of the spatial heterogeneity of resources and fishing, especially in a fragmented coral reef habitat.

Among existing spatial fisheries models, the ISIS-Fish model [1] was designed to be parameterized from existing knowledge on fish resources and corresponding fisheries. In the present paper, we used the ISIS-Fish tool to build a spatially-explicit dynamic model for two major fish resources: spangled emperor (*Lethrinus nebulosus*, Forsskal, 1775) and coral trout (*Plectropomus leopardus*, Lacepède, 1802). After calibration, it was used to simulate the outcomes of several management scenarios including MPA, size limitation, and increase fishing effort (number of commercial fishermen).

2   Material and Methods

ISIS-Fish relies on a grid of identical cells from which model zones are defined. The extent of the modelled area was defined to include the known features of population dynamics of the species studied. This large area encompasses the southwest lagoon of New Caledonia from fringing reef to barrier reef. Cell size was defined as a trade-off between a) the ability to identify some important features of habitat and small existing marine reserves and b) computational constraints, and resulting in a square grid side of 0.01 degree.

For mobile species, modeling population dynamics requires to account and formalize mobility. Mobility coefficients were defined to characterise fish movement between model zones, with regard to the spawning season and dispersion. Spangled emperor is mobile and macrocarnivore. It exhibits a strong preference for sea grass and sandy bottom habitats, and it changes habitat according to ontogeny, while coral trout is piscivore, mainly sedentary and preferably found on coral reef habitats (see references in [2]). The two resources are targeted by distinct fishing métiers, spangled emperor being caught by line fishing, while coral trout is mostly caught by spearfishers.

Professional and recreational fishing were characterised using respectively fisheries data and interview data.

The model was calibrated from existing abundance indices obtained from underwater visual counts using the Simplex algorithm.

Sensitivity analysis was achieved using the group-screening technique [3] within a fractional design [4]. A total of 140 parameters were identified as uncertainty-prone, 77 for population submodel and 63 for fishery submodel. These parameters were clustered into 18 groups bearing similar consequences on model outputs. For each parameter, the range of variation considered was either the minimum and maximum values found in the literature, or alternatively a 20% range around the nominal parameter value.

3   Results

Simulation results showed that the consequences of alternative scenarios on population dynamics could not all be distinguished, in relation with parameter uncertainty. Under certain assumptions and depending on the species, the proposed MPA design could result in increased population abundances within its boundaries. The minimum legal size scenario strongly reduced line fishing catches, and spangled emperor and coral trout biomass only benefitted from the latter scenario. But outcomes depended on the survival rate of released fish. Increasing the number of professional fishers significantly affected spangled emperor populations, but not coral trout. Other MPA scenario could be investigated from the model.

Uncertainties on coral trout recruitment resulted in contrasted biomass trajectories, depending on the assumption, whatever the management scenario.

This work shows that, from existing data, it was possible to parameterize a complex spatial model for two major fished resources. The model enabled to confront the consequences of a range of plausible management scenarios. These consequences depend on existing uncertainties, some of which are tied to poorly-quantified aspects of population dynamics. The study thus points out critical gaps in resource knowledge which need to be addressed to better appraise population dynamics under a range of management scenarios.

References

1. Pelletier, D., Mahévas, S., Drouineau, H., Vermard, Y., Thébaud, O., Guyader, O., Poussin, B.: Evaluation of the bioeconomic sustainability of complex fisheries under a wide range of policy options using ISIS-Fish. Ecological Modelling 220, 1013–1033, doi:10.1016/j.ecolmodel.2009.01.007
2. Preuss, B.: Évaluation de scénarios de gestion des ressources du lagon Sud-ouest de la Nouvelle-Calédonie: Intégration des connaissances et modélisation spatialement explicite. PhD dissertation. Université de Nouvelle-Calédonie, 386 p. (2012)
3. Satelli, A.: What is sensitivity analysis? In: Saltelli, A., Chan, K., Scott, E.M. (eds.) Sensitivity Analysis, Probability and Statistics, pp. 3–13. John Wiley, New York (2004)
4. Droesbeke, J., Fine, J., Saporta, G.: Plans d' Expérience. Applications á l' Entreprise. Technip, Paris (1997)

Completing SBGN-AF Networks
by Logic-Based Hypothesis Finding

Yoshitaka Yamamoto[1], Adrien Rougny[2], Hidetomo Nabeshima[1],
Katsumi Inoue[3], Hisao Moriya[4], Christine Froidevaux[2], and Koji Iwanuma[1]

[1] University of Yamanashi
4-3-11 Takeda, Kofu-shi, Yamanashi 400-8511, Japan
[2] Laboratoire de Recherche en Informatique (LRI), CNRS UMR 8623
Université Paris Sud, France
[3] National Institute of Informatics
2-1-2 Hitotsubashi, Chiyoda-ku, Tokyo 101-8430, Japan
[4] RCIS, Okayama University
3-1-1 Tsushimanaka, Kita-ku, Okayama 700-8530, Japan

Abstract. This study considers formal methods for finding unknown interactions of incomplete molecular networks using microarray profiles. In systems biology, a challenging problem lies in the growing scale and complexity of molecular networks. Along with high-throughput experimental tools, it is not straightforward to reconstruct huge and complicated networks using observed data by hand. Thus, we address the completion problem of our target networks represented by a standard markup language, called SBGN (in particular, Activity Flow). Our proposed method is based on logic-based hypothesis finding techniques; given an input SBGN network and its profile data, missing interactions can be logically generated as hypotheses by the proposed method. In this paper, we also show empirical results that demonstrate how the proposed method works with a real network involved in the glucose repression of *S. cerevisiae*.

Keywords: completion, hypothesis finding, SBGN, glucose repression.

1 Introduction

Systems biology has been developed to reconstruct complex and diverse cellular mechanisms involved in genome, proteome and metabolome into a whole system. Those mechanisms consist of molecular interactions or causalities, which are often represented as a network, called a *molecular network*. In the last decade, formal languages have been developed for describing molecular networks in some standardized graphical form. One of the main standards is the *Systems Biology Graphical Notation* (SBGN) [12]. SBGN enables us to represent any individual molecular network in a uniformed and shareable way.

There are recent works [10, 11] to logically formalize the semantics of SBGN Activity Flow (SBGN-AF) and translate it into first-logic logic [10] or normal

F. Fages and C. Piazza (Eds.): FMMB 2014, LNBI 8738, pp. 165–179, 2014.
© Springer International Publishing Switzerland 2014

logic programming [11]. Such translation allows us to analyze SBGN-AF networks using existing inference engines with richer knowledge representation formalisms. Rougny *et al* [10] have used their translation for analyzing the dynamics of SBGN networks with a deductive method like SOLAR [17]. Unlike their previous work, this study addresses the *completion problem* to find missing interactions from SBGN networks together with their microarray profiling (i.e., gene expression) data. Along with high-throughput experimental tools, the whole network rapidly becomes larger and more complicated, which will become difficult to be managed by hand. In turn, the observed data also becomes larger and more comprehensive. Thus, it is required to automatically infer possible missing interactions in the prior network so as to fit observations.

There are related works dealing with the completion problem [1, 3, 4, 6, 8, 9, 18, 20, 21, 23, 24]. In terms of Boolean networks, the computational difficulty has been investigated in [1]. This work focuses on the problem to determine Boolean functions for unassigned nodes with a given set of examples, and shows that it is NP-hard in the general case. In order to cope with this computational intractability, various methods with heuristics have been proposed [18]. Karlebach and Shamir [8] have proposed an entry based approach without discretizing observed data. From the viewpoint of applying inference techniques, the literature [20] has focused on transcriptional-regulatory networks (TRNs) and investigated the previously proposed inferring techniques for TRNs. The literature [3, 4, 21, 23, 24] shows that logic-based hypothesis finding techniques enable us to find unknown effects or interactions in real biological networks, like metabolic networks and signaling networks. Kleinberg and Mishra [9] have proposed a framework for inferring causal structures using the temporal logic from time-course data. Inoue *et al* [6] formalized biological networks as so-called *causal networks* and proposed the methodology, called *meta-level abduction*, for finding missing causal relations in the prior causal network. In contrast, this study treats the completion problem of SBGN networks using logic-based hypothesis finding techniques, which have been mainly developed in Inductive Logic Programming (ILP) [16, 19]. In ILP, the task of hypothesis finding is to find a consistent *hypothesis H* such that H logically explains a given observation E with respect to the prior background theory B. Note that B, E and H correspond to the prior input of SBGN network, its microarray profiling data and missing interactions, respectively. Thus, it is necessary to formalize SBGN and profiling data into B and E that are given as clausal theories.

Based on the previously proposed translation techniques [6, 10], we focus on SBGN-AF language, which is best used for representing gene regulations and signaling networks. Given a SBGN-AF network, we formalize its topological knowledge using two predicates: *stimulates*/2 and *inhibits*/2, which mean the positive and negative regulation between nodes, respectively. The profiling data is generated by collecting gene expression changes, called *fold changes*, caused by some knockout perturbation. We discretize these fold change values in accordance with some threshold and p-values into their change states (either *up*, *down*, or *stable*). This knowledge is represented using two predicates: *promotes*/2 and *suppresses*/2 for the knockout perturbation *ko*. If the expression level of some gene is changed by *ko*, there should exist a cascade of positive or negative regulations from *ko* to the gene. In [6], causal relations between such chain effects and direct regulations have been carefully investigated. These relations allow us to realize a "gap" between the observed data (i.e., chain effects) and the prior SBGN-AF network (i.e., direct regulations), which is logically derivable in the context of ILP.

In this paper, we demonstrate how the proposed method works in the completion problem of SBGN-AF networks using the *glucose repression* system in the yeast *S. cerevisiae* [2, 22]. Glucose repression has evolved as a complex regulatory system consisting of several different pathways. Recently, they have been reconstructed into a logical hypergraph that can be represented in SBGN-AF, and evaluated by comparing changes in the logical state of gene nodes with several kinds of profiling data. However, the percentage of true predictions remains about 50% on average, and thus the current network is too incomplete to predict the gene expressions correctly. The experimental result shows that the proposed logic-based method can partially succeed in finding missing interactions whose relevance can be confirmed in the YEASTRACT database[1] for yeast.

The rest of this paper is organized as follows. Section 2 introduces the notions and terminology used in the paper and briefly reviews the techniques for hypothesis finding. Section 3 introduces the motivating example on glucose repression and formalizes the SBGN completion problem for it. In Section 4, we show the experimental result obtained by using the motivating example. We then conclude in Section 6.

2   Background

2.1   Inductive Logic Programming

Here, we review the notation and terminology used in ILP [19]. A *clause* is a finite disjunction of literals which is often identified with the set of its literals. A clause $\{A_1, \ldots, A_n, \neg B_1, \ldots, \neg B_m\}$, where each A_i, B_j is an atom, is also written as $A_1 \vee \cdots \vee A_n \leftarrow B_1 \wedge \cdots \wedge B_m$. A *positive* (*negative*) *clause* is a clause whose disjuncts are all positive (negative) literals. A *unit clause* is a clause with exactly one literal. A *clausal theory* is a finite set of clauses. A clausal theory Σ is often identified with the conjunction of its clauses and is said to be in *Conjunctive Normal Form* (CNF). Let Σ_1 and Σ_2 be two clausal theories. Σ_1 and Σ_2 are said to be equivalent, denoted $\Sigma_1 \equiv \Sigma_2$, if $\Sigma_1 \models \Sigma_2$ and $\Sigma_2 \models \Sigma_1$, where $\models$ means the standard entailment relation. For a clausal theory Σ, a *consequence* of Σ is a clause entailed by Σ. We denote by $Th(\Sigma)$ the set of all consequences of Σ. We give the definition of hypotheses as follows:

Definition 1 (Hypothesis). Let B and E be clausal theories, representing a background theory and (positive) examples/observations, respectively. Then H

---

[1] Available from http://www.yeastract.com/

is a *hypothesis wrt B and E* iff H is a clausal theory such that $B \land H \models E$ and $B \land H$ is consistent. If no confusion arises, then we refer simply to a "hypothesis" instead of a "hypothesis wrt B and E".

2.2 Systems Biology Graphical Notation (SBGN)

SBGN [12] is one of the main standardized formalization for representing molecular networks, and used in reaction databases like Reactome. It consists of three languages: Process Description (SBGN-PD), Activity Flow (SBGN-AF) and Entity Relationship (SBGN-ER). Each of them describes its own biological aspect. In the following, we focus on the SGBN-AF class which is used to represent biological activities and their influences on each other [10, 14]. It is best suited for representing gene regulations and signaling networks. The SBGN-AF language contains a set of glyphs, which can be classified into five groups: activity nodes, auxiliary units, container nodes, modulation arcs and logical operators. In [10, 11], those SBGN-AF glyphs have been interpreted using predicate logic. For example, five kinds of glyphs in modulation arcs termed by positive influence, negative influence, unknown influence, necessary stimulation and logic arc are interpreted with the predicate *stimulates*/2, *inhibits*/2, *unknownInfluences*/2, *necessaryStimulates*/2 and *input*/2, respectively.

Example 1. Figure 1 describes a simple example of SBGN-AF network. The translation of this network is achieved by using the following unit clauses:

Activity nodes : $activity(a_1)$. $activity(a_2)$. $activity(a_3)$. $activity(a_4)$.
Logical operators : $or(lo_1)$. $and(lo_2)$. $not(lo3)$.
Modulation arcs : $input(a_1, lo_1)$. $input(a_2, lo_1)$. $input(lo_1, lo_2)$. $input(lo_3, lo_2)$.
    $stimulates(lo_2, a_4)$. $inhibits(a_4, a_1)$.

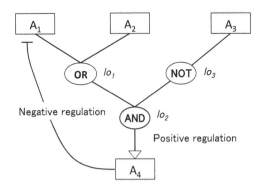

Fig. 1. Example of SBGN-AF network

If each of four nodes has a Boolean state, this network may be interpreted as the Boolean network consisting of two functions $a_4 = (a_1 \lor a_2) \land \neg a_3$ and $a_1 = \neg a_4$.

Note that SBGN-AF can represent any Boolean network, since every Boolean function of form: $y = f(x_1, x_2, \ldots, x_n)$ is drawn with $n+1$ activity nodes, logical operators, positive and negative influence arcs.

2.3   SBGN-AF Networks as Causal Networks

Inoue *et al* [6] have proposed the notion of a *causal network* representing causal relations, which consists of a set of nodes and a set of directed arcs. Each node denotes an *event*, *fact* or *proposition* and each directed arc represents a *directed causal relation*, which corresponds to a causality between nodes. In the literature, the interpretation of causalities is kept informal, and just represents the connectivity, which may refer to a mathematical, physical, chemical, conceptual, epidemiological, structural, or statistical dependency [6]. In this study, we treat molecular interactions in SBGN-AF as causal dependencies, which can be evaluated with biological evidences observed by knock-out experiments. One may formally give the interpretation of those molecular causalities in previously proposed frameworks such as *Lewis's counterfactual theory* [13, 15].

In the context of SBGN-AF, causal networks capture such SBGN-AF networks that are represented with activity nodes and positive and negative regulation arcs as well as three logical operators for "and", "or" and "not". Unlike [10], formalization of causal networks[2] is achieved by expressing causal relations between objects at the meta level using so-called *meta-predicates*: *stimulates*/2 and *inhibits*/2. In a biological viewpoint, these two atoms represent that for two activity nodes a and b, the steady state of b is significantly changed by some perturbation (i.e., knock-out effect) to a. Suppose that if a is knocked out, the expression level of b decreases. This fact indicates that there is a positive influence from a to b, which is represented as *stimulates*(a, b). In contrast, if the expression level of b conversely increases, then we recognize that there is a negative influence from a to b, which is represented as *inhibits*(a, b). In this paper, an atom *stimulates*(a, b) (*resp. inhibits*(a, b)) at the meta level is logically interpreted as the causal relation $b \leftarrow a$ (*resp.* $b \vee a$) at the object level. This logical interpretation enables us to implicitly represent the logical operators, as shown in the below example.

Example 2. We recall Example 1. At the object level, the causal relation between four nodes is represented by the two logical formulas $a_4 \leftarrow ((a_1 \vee a_2) \wedge \neg a_3)$ and $a_1 \vee a_4$. The former formula is written as the clausal form of $(a_4 \vee a_3 \leftarrow a_1) \wedge (a_4 \vee a_3 \leftarrow a_2)$. We can rewrite this as $((a_4 \leftarrow a_1) \vee (a_4 \vee a_3)) \wedge ((a_4 \leftarrow a_2) \vee (a_4 \vee a_3))$. Then, using the meta-predicates, this causal relation is represented by the two clauses: *stimulates*$(a_1, a_4) \vee inhibits(a_3, a_4)$ and *stimulates*$(a_2, a_4) \vee inhibits(a_3, a_4)$. We also have *inhibits*(a_4, a_1) from the latter formula $a_1 \vee a_4$.

The SBGN-AF translation [10, 11] needs to explicitly describe the intermediate nodes for logical operators, whereas the obtained theory is uniquely determined.

---

[2] In the original setting, they use two predicates: *triggered*/2 and *inhibited*/2, which can be regarded as *stimulates*/2 and *inhibits*/2, respectively, in this paper.

In contrast, formalization of causal networks is simply given as the clausal form only with the two meta-predicates: *stimulates* and *inhibits*, whereas the way for formalizing networks can be indeterminate. Indeed, there are multiple ways to write a clause as the implication form in first-order logic. For instance, the clause $\{a, b, \neg c\}$ is written as either $(a \vee b) \wedge (b \leftarrow c)$, $(a \vee b) \wedge (a \leftarrow c)$ or $(a \leftarrow c) \wedge (b \leftarrow c)$. Recall Example 2. The causal relation $a_4 \leftarrow ((a_1 \vee a_2) \wedge \neg a_3)$ can be alternatively written as $((a_4 \leftarrow a_1) \vee (a_3 \leftarrow a_1)) \wedge ((a_4 \leftarrow a_2) \vee (a_3 \leftarrow a_2))$. In this case, the causal relation is represented by other two clauses: $stimulates(a_1, a_4) \vee stimulates(a_1, a_3)$ and $stimulates(a_2, a_4) \vee stimulates(a_2, a_3)$. In this paper, we fix the head atom in each implication form as the one in an original causal relation to copy with the ambiguity in formalizing causal networks. Note that the head atom is a_4 in the above example. If we fix the head atom as a_4 in each implication form, we have the unique translated formulas in Example 2.

We also remark that this study focuses on *gene expression changes in the steady state*, rather than time-series expression changes. Note that if we need to treat time-series data, it is sufficient to add a new term corresponding to the *time* concept in the prior atoms.

2.4   Meta-Level Abduction in Causal Networks

Next, we introduce two predicates *suppresses*/2 and *promotes*/2 for representing causal chains [6]. Figure 2 describes the causal relation between chain

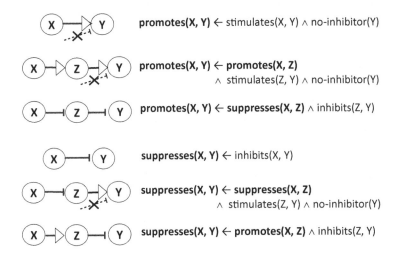

Fig. 2. Causalities between chain regulations and direct regulations

regulations and directed positive and negative regulations. Note that an atom *no_inhibitor(Y)* corresponds to the default assumption that there is no negative regulation to Y. In other words, we assume that any positive regulation to Y works only in case that Y is not involved in any negative regulation.

Suppose that chain regulations are given as observations E. Then, the above causal rules allow us to detect unknown direct regulations as a hypothesis H such that H logically explains E with the prior background theory B. Hence, B contains the topological knowledge of the network and the causal rules in Fig. 2. Hypotheses can be generated by using the consequence finding system SOLAR [17], which is based on SOL resolution [5] and the connection tableaux [7].

SOLAR efficiently computes so-called *characteristic clauses* which represent "interesting" consequences of a given problem for users. Each characteristic clause is constructed over a sub-vocabulary of the representation language called a *production field*. A production field $\mathcal{P}$ is defined as a pair, $\langle \mathbf{L}, Cond \rangle$, where $\mathbf{L}$ is a set of literals closed under instantiation, and $Cond$ is a certain condition to be satisfied, e.g., the maximum length of clauses, the maximum depth of terms, etc. When $Cond$ is not specified, $\mathcal{P}$ is simply denoted as $\langle \mathbf{L} \rangle$. A clause C *belongs* to $\mathcal{P} = \langle \mathbf{L}, Cond \rangle$ if every literal in C belongs to $\mathbf{L}$ and C satisfies $Cond$. For a set Σ of clauses, the set of consequences of Σ belonging to $\mathcal{P}$ is denoted $Th_{\mathcal{P}}(\Sigma)$. Then, the characteristic clauses of Σ wrt $\mathcal{P}$ are defined as: $Carc(\Sigma, \mathcal{P}) = \mu Th_{\mathcal{P}}(\Sigma)$, where μT denotes the set of clauses in T that are minimal with respect to subsumption. When a new clause F is added to a clausal theory, some consequences are newly derived with this additional information. The set of such clauses that belong to the production field are called *new characteristic clauses*. Formally, the *new characteristic clauses* of F wrt Σ and $\mathcal{P}$ are defined as $NewCarc(\Sigma, F, \mathcal{P}) = Carc(\Sigma \cup \{F\}, \mathcal{P}) - Carc(\Sigma, \mathcal{P})$.

Given an observation E consisting of one chain regulation and the background theory B, we seek for a hypothesis H wrt B and E that consists of direct regulations. By the principle of inverse entailment [16], it holds that $B \wedge \neg E \models \neg H$ and $B \not\models \neg H$. Thus, $\neg H$ is regarded as a consequence of $B \wedge \neg E$ belongs to $\mathcal{P}$ which is not a consequence of $\neg E$ only. Hence, the set of minimal hypotheses wrt B, E and $\mathcal{P}$ is characterized as $\{H \mid \neg H \in Newcarc(B, E, \mathcal{P})\}$. Suppose that the abducibles $\mathbf{L}$ where any literal in H must be an instance of a literal are given. Then, it is sufficient to set the production field $\mathcal{P}$ as $\langle \overline{\mathbf{L}} \rangle$ where $\overline{\mathbf{L}}$ is the set $\{\neg l \mid l \in \mathbf{L}\}$. Note that $\mathbf{L}$ for meta-level abduction is given as the possible missing arcs represented by two predicates *stimulates* and *inhibit* as well as the predicate for default assumption *no_inhibitor*.

3 Motivating Example

We empirically show that our formal method for solving the SBGN-AF completion problem is applicable in real biological systems.

3.1 Glucose Repression

Now, we explain the glucose repression system in yeast *S. cerevisiae* [22]. This is an essential system that most eukaryotic cells including yeasts and humans have. By this system, cells are used to sense the availability of carbon sources and if there is their favorite sugar like glucose, then they repress the utilization of less-favorite sugars. It is known that this system is achieved by extensively cross-talking two kinds of pathways for sensing and repression (See Figure 3). In the left-hand side of Figure 3, sensing proteins on the cell membrane inhibit the expressions of both Mth1 and Std1, which control the transcriptional factor $Rgt1p$ that regulates the expression level of transporting proteins Hxts. In case that there is glucose, transporting proteins are activated and then allow glucose to move inside the cell. Thus, glucose is used for generating energies in biological process, called *fermentation*. On the other hand, if there is no glucose, transporting proteins are not expressed. In this case, cells use less-favorite sugars; otherwise ethanol is used to generate energies in so-called *oxidation* process. Eukaryotic cells have such mechanisms that dynamically change their activated pathways in accordance with the existence of favorite sugars (i.e., glucose).

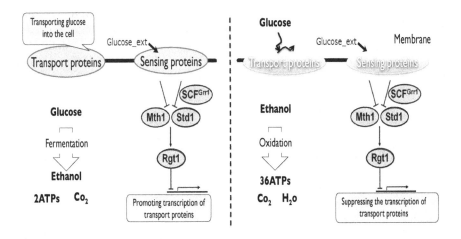

Fig. 3. Dynamics of the glucose repression system [22]

In date, the whole mechanism has been reconstructed as one hypergraph that consists of 83 nodes and 118 hyperedges [2]. This hypergraph can be represented with SBGN-AF, and is regarded as a causal network. Thus, we can consider the completion problem of this SBGN-AF network together with observed data.

In this paper, we utilize the microarray profiling data which consists of each gene expression in wild type (WT) and the mutation $(\triangle X)$ obtained by knocking out the gene X. We first obtain the fold change of each gene by comparing its (raw) expression data in WT with the one in $\triangle X$. Next, we discretize this change value into the 3 values (*up*, *down*, and *stable*) using some threshold. In this paper, we set this threshold to 1.5, as this value is used as default in the literature [2]. Let x be a fold change value. If $x \geq 1.5$, then the change state is *up*, else if $-1.5 < x < 1.5$, the state is *stable*, otherwise it is *down*. Note that the change state of a gene is regarded as the causal chain from the knockout effect ko to the gene. Thus, we formalize *up*, *down* and *stable* state of a gene g with the atom $promotes(ko, g)$, $no\_changes(ko, g)$ and $suppresses(ko, g)$, respectively.

Table 1. An example for discretizing gene expression data

gene	WT	$\triangle X$	fold change	change state	observation
X	200	100	-2.0	*down*	*suppresses(ko, X)*
Y	100	180	1.8	*up*	*promotes(ko, Y)*
Z	100	110	1.1	*stable*	*no_changes(ko, Z)*

Christensen *et al* [2] have used the microarray data with respect to the yeast mutations $\triangle RGT1$, $\triangle MIG1$, $\triangle MIG1\triangle MIG2$, $\triangle SNF1\triangle SNF4$, $\triangle GRR1$, and show that the prior model is able to predict each gene expression change with about 50% accuracy on average. This result however indicates the existence of unknown causalities on gene/protein regulations that are missed in the current model. In this paper, we use three types of microarray data on $\triangle MIG1$, $\triangle MIG1\triangle MIG2$ and $\triangle GRR1$ that are available so far. By discretizing them with the threshold 1.5, we obtain 15, 20, and 28 chain causalities on $\triangle MIG1$, $\triangle MIG1\triangle MIG2$, and $\triangle GRR1$, respectively. Then, we empirically investigate if the proposed method can generate hypotheses for each of those observed facts.

Note that for each gene, there are candidates of transcription factors (TFs) that can bind it. These possible TFs are divided into two groups: documented and potential ones. The former group has empirical proofs in the literature, whereas the latter one is detected from their base sequence using bioinformatics. Both data are stored in the yeast database YEASTRACT. By crawling the database, we obtain the 149 pairs of each gene and its TFs in total. For each pair $\langle Gene, TF \rangle$, there are two possible regulations: positive and negative ones, which can be represented as two atoms $stimulates(TF, Gene)$ and $inhibits(TF, Gene)$, respectively. The hypotheses to be generated is then composed from those negated literals (i.e., 298 negative literals). Finally, we put them in the abducibles **L** of the production field $\mathcal{P}$.

3.2 Solving a Toy SBGN-AF Completion Problem

The prior glucose repression model can be regarded as a causal network. Then, we formalize the topology based on the meta-level translation [6] and also use the causal rules in Fig. 2 for describing chain regulations. Both of the topology and causal rules are contained in the background theory B.

Given B, E and $\mathcal{P}$, we can compute the minimal hypotheses wrt B, E and $\mathcal{P}$ by using SOLAR. The obtained hypothesis corresponds to the set of missing transcription factors that have been stored in YEASTRACT. Here, we demonstrate how our logic-based method can generate hypotheses using a toy example. In Figure 4, we treat only one direct regulation involving two proteins $Mig1p$, $Snf1p$ and one gene $MIG1$. Suppose that the expression of $YGL157W$ is promoted by knocking out the gene $MIG1$, which is written as $promotes(ko, ygl157w)$. For this observation E, we can consider one possible interaction between $Mig1p$ and $YGL157W$, which is lacked in the prior network but documented in the database. Fig. 4 describes the prior network and the observation. Given E, we now derive the above hypothesis H meaning that the transcription factor $Mig1p$ suppresses $YGL157W$ using the background theory B consisting of the following:

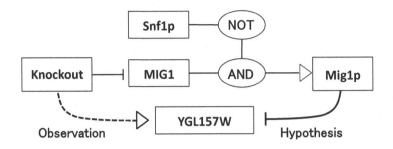

Fig. 4. Simplified network involving $MIG1$ and $YGL157W$

1. **Knockout perturbation** : $suppresses(ko, mig1)$.
2. **Topology** : $stimulates(mig1, mig1p) \lor inhibits(snf1p, mig1p)$.
3. **Causalities** : $promotes(X, Y) \leftarrow inhibits(Z, Y) \land suppresses(X, Z)$.
   $suppresses(X, Y) \leftarrow stimulates(Z, Y) \land suppresses(X, Z) \land no\_inhibitor(Y)$.
4. **Integrity constraint** $\leftarrow no\_inhibitor(X) \land inhibits(Y, X)$.

There are several remarks as follows:

1. We give the direct effect by knocking out $MIG1$ in the background theory.
2. The interaction between $MIG1$, $Mig1p$ and $Snf1p$ is originally represented as the causal relation $mig1p \leftarrow mig1 \land \neg snf1p$. This is written as the causal form $(mig1p \leftarrow mig1) \lor (mig1p \lor snf1p)$. We rewrite it using two meta-level predicates: $stimulates(mig1, mig1p)$ and $inhibits(snf1p, mig1p)$.
3. We use two causal rules on the meta-predicate *promotes* and *suppresses*. Note that the predicate $no\_inhibitor/1$ is used to assume that the target gene is not inhibited from any implicit factor. It is thus treated as a default. It is necessary to put the negation of this default atom for realizing default reasoning in SOLAR[3].
4. The integrity constraint on *no_inhibitor* is introduced. If there is some explicit factor inhibiting the gene X, then $no\_inhibitor(X)$ is no longer true.

We next give the production field $\mathcal{P}$ as $\langle \mathbf{L} \rangle$ where $\mathbf{L}$ is as follows:

$$\mathbf{L} = \{ \ \neg no\_inhibitor(mig1p),$$
$$\neg stimulates(mig1p, ygl157w), \neg inhibits(mig1p, ygl157w),$$
$$\neg stimulates(snp1p, ygl157w), \neg inhibits(snp1p, ygl157w) \ \}$$

Together with B, E and $\mathcal{P}$, we can derive the following new characteristic clause C of $\neg E$ wrt B and $\mathcal{P}$ by SOLAR:

$$C = \neg inhibits(mig1p, ygl157w) \lor \neg no\_inhibitor(mig1p).$$

---

[3] We refer to [6] for more detail.

We may notice that the target hypothesis H corresponds to $\neg C$. Indeed, it holds that $B \wedge \neg C \models E$ and $B \wedge \neg C$ is consistent. Note here that there is no ORF name for the gene $YGL1157W$, which means its particular functionality has not been clarified yet. Indeed, the prior network does not contain the inhibitory relation between $Mig1p$ and $YGL157W$, though we can detect one missing transcriptional interaction by the proposed method.

Computation of new characteristic clauses is based on the Skip Ordered Linear (SOL) resolution [5], which realizes an efficient calculus for consequence finding. SOLAR [17] is a sophisticated implementation for SOL calculus with various kinds of pruning methods [7]. SOLAR computes each new characteristic clause of $\neg E$ by generating so-called *SOL tableaux* with four kinds of operations: *extension, skip, reduction, skip-factoring*. Figure 5 describes the SOL tableaux for deriving the target clause C from $\neg E$. In the figure, we abbreviate the predicate name *promotes, suppresses, stimulates, inhibits* and *no_inhibitor* to *pro, sup, sti, inh, no-inh*, respectively. We also denote by the term y the gene $YGL157W$. The root node is labeled with the clause $\neg E$. Then, SOLAR applies the extension operator to concatenate two complementary literals $pro(ko, y)$ of $\neg E$ and $\neg pro(ko, y)$ of a ground clause in B. This operation enables us to derive a consequence $\{\neg inh(mig1p, y), \neg sup(ko, mig1p)\}$. Since the literal $\neg inh(mig1p, y)$ belongs to the production field $\mathcal{P}$, we can terminate the derivation step here, which is called *skip* operation. We next continue the extension operation to the other literal $\neg sup(ko, mig1p)$. Finally, we obtain a new characteristic clause consisting of two skipped literals: $\neg inh(mig1p, y)$ and $\neg no\text{-}inh(mig1p)$.

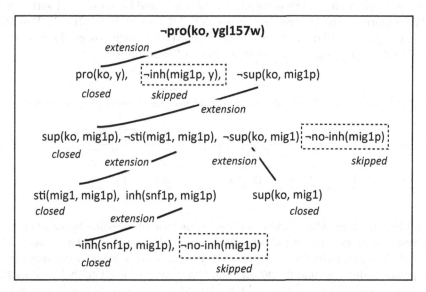

Fig. 5. SOL tableaux generated by SOLAR

The SOL calculus is sound and complete for finding the new characteristic clauses, although the search space is huge due to the nondeterminisms for applying possible operations. Thus, it is necessary to prepare for a sufficient production field in the practical viewpoint of SOLAR. In this paper, the experiment aims at seeking hypotheses for the possible transcription factors stored in YEASTRACT. Then, we assume the production field reflecting those possible transcription factors each of which can bind the target gene.

4   Experiments

Next, we investigate whether or not our proposed method can derive sufficient hypotheses from each observation for $\triangle MIG1$, $\triangle MIG1 \triangle MIG2$ and $\triangle GRR1$ using the background theory, which has been described in the previous sections. For the experiment, we provide three types of production field as follows:

1. $\mathcal{P}_1 = \langle \mathbf{L_1} \rangle$ where $\mathbf{L_1}$ consists of only the atom for default reasoning (i.e., $\mathbf{L_1} = \{\neg no\_inhibitor(X)\}$).
2. $\mathcal{P}_2 = \langle \mathbf{L_2} \rangle$ where $\mathbf{L_2}$ consists of $\mathbf{L_1}$ and the set of atoms for documented transcriptional regulations stored in the database.
3. $\mathcal{P}_3 = \langle \mathbf{L_3} \rangle$ where $\mathbf{L_3}$ consists of $\mathbf{L_2}$ and the set of atoms for potential transcriptional regulations stored in the database.

If SOLAR can derive some hypothesis only with $\mathcal{P}_1$ for some observation, we do not need to augment the prior network using any new regulation. In contrast, such hypotheses that can be derived with $\mathcal{P}_2$ have been already proved in a biological viewpoint, and thus the prior network should be augmented with them.

We performed the experiments using SOLAR (ver. 2, build 314) in Red Hat Linux machines (CPU: 2.67GHz, Memory: 8GM). For each observed fact in each profiling data, we limit the execution time to 100 seconds.

Table 2. The number of observations that succeed in generating hypotheses

production field	$\triangle MIG1$	$\triangle MIG1 \triangle MIG2$	$\triangle GRR1$
$\mathcal{P}_1$	2 (15)	5 (20)	0 (28)
$\mathcal{P}_2$	15 (15)	20 (20)	4 (28)
$\mathcal{P}_3$	15 (15)	20 (20)	4 (28)

Table 2 describes the number of observations for which some hypotheses were derived with respect to each observed fact and each type of production field. The number shown in parentheses denotes the number of observations obtained from the profiling data of each mutation. The experimental result with $\mathcal{P}_1$ shows the number of observations that can be explained by the prior network, which corresponds to two and five for $\triangle MIG1$ and $\triangle MIG1 \triangle MIG2$, respectively. Thus, the predictive accuracy based on those numbers are significantly lower than the previously reported values in [2], although the methodology for computing the accuracy is different from each other.

In this experiment, for every observation on $\triangle MIG1$ and $\triangle MIG1\triangle MIG2$, we could derive some hypotheses using the production field $\mathcal{P}_2$. Accordingly, the observations on $\triangle MIG1$ and $\triangle MIG1\triangle MIG2$ can be also explained with $\mathcal{P}_3$, since $\mathcal{P}_3$ includes $\mathcal{P}_2$. In contrast, we could derive hypotheses only for four observations ($HXT1$, $HXT3$, $HXT4$, $YGL157W$) on $\triangle GRR1$, even if we use the production field $\mathcal{P}_3$. We have performed another experiment where the executing time was limited to 300 seconds, although the result was the same as Table 2.

To improve the performance of SOLAR on $\triangle GRR1$, we have tried a framework for incremental hypothesis finding. Note that the observations E are given as the conjunction of observations each of which is represented by using either *promotes* or *suppresses* atom. So far, we have computed $NewCarc(B, \neg e, \mathcal{P})$ for each observation e in E in a divide and conquer manner of SOLAR. Once the current observation e was explained, we can treat it in the background theory to compute the hypothesis for the next observation. In other words, we can reuse the previously proved observations as *lemma*, which may be similar to the notion of *Splitting Lemmas* in MGTP. We have performed the exper-

---

Begin
While (true)
    For each e in E
       $Se := NewCarc(B, \neg e, \mathcal{P})$;
       If Se is not empty, then add e to the set F and remove e from E;
       If E or F is empty, then break;
       Else, add F to B and reset F to the empty set;
End

Fig. 6. Sketch of incremental hypothesis fining with lemmas

iment using this incremental hypothesis finding with lemmas, as sketched in Figure 6, although the result was same as the previous one in case that the execution time for computing $NewCarc(B, \neg e, \mathcal{P})$ is limited to 300 seconds. Then, we next checked whether or not SOLAR can compute $NewCarc(B \cup \{E - \{e\}\}, \neg e, \mathcal{P})$ for each observation e within 300 seconds as a preliminary experiment. Consequently, SOLAR could succeed in deriving hypotheses for most observations in this setting. Based on this result, one may consider that SOLAR has failed to prove some intermediate lemma that was difficult to be proved but necessary to compute consequences. Thus, we intend to furthermore investigate the inference process in SOLAR and improve the knowledge representation of B in order to to solve the observations on $\triangle GRR1$.

5   Conclusion and Future Work

In this paper, we have addressed the completion problem of SBGN-AF networks using the logic-based hypothesis finding method. We have applied our proposed method to the *glucose repression* network in the yeast *S. cerevisiae*. The glucose repression has evolved as a complex regulatory system consisting of several different pathways. For this real system, we have empirically shown that the logic-based method can partially succeed in finding missing interactions, which have been confirmed in the database. As shown in the toy example, it is difficult to logically detect possible interactions even in such a small network with one interaction. Along with high-throughput experimental tools, the whole network rapidly becomes larger and more complicated. In this context, it will be more required to automatically find possible missing interactions in the prior network so as to fit observations. Thus, one of the most important issues is to investigate the applicability of our logic-based method to other real networks whose scales are at least comparable with the glucose repression system.

Acknowledgement. This work was supported by NII Collaborative Research Project 2013 on "Advanced Reasoning for Analyzing Molecular Networks", and also supported by JSPS KAKENHI Grant Number 25730133. The authors would like to thank the anonymous reviewers for giving us useful and constructive comments.

References

1. Akutsu, T., Tamura, T., Horimoto, K.: Completing networks using observed data. In: Gavaldà, R., Lugosi, G., Zeugmann, T., Zilles, S. (eds.) ALT 2009. LNCS (LNAI), vol. 5809, pp. 126–140. Springer, Heidelberg (2009)
2. Christensen, T.S., Oliveira, A.P., Nielsen, J.: Reconstruction and logical modeling of glucose repression signaling pathways in *Saccharomyces cerevisiae*. BMC Systems Biology 3 (2009), doi:10.1186/1752-0509-3-7
3. Demolombe, R., Farinas del Cerro, L., Obeid, N.: A logical model for metabolic networks with inhibition. In: Proc. of Int. Conf. on Bioinformatics and Computational Biology (2013)
4. Doncescu, A., Inoue, K., Pradine, A.: MicroRNA analysis by hypothesis finding technique. In: Late breaking Papers from the 22nd Int. Conf. on Inductive Logic Programming (ILP 2012), CEUR, vol. 975, pp. 26–37 (2013)
5. Inoue, K.: Linear resolution for consequence finding. Artificial Intelligence 56, 301–353 (1992)
6. Inoue, K., Doncescu, A., Nabeshima, H.: Completing causal networks by meta-level abduction. Machine Learning 91, 239–277 (2013)
7. Iwanuma, K., Inoue, K., Satoh, K.: Completeness of pruning methods for consequence finding procedure SOL. In: Proc. of the 3rd worksh. on First-order Theorem Proving (FTP 2000), pp. 89–100 (2000)
8. Karlebach, G., Shamir, R.: Constructing logical models of gene regulatory networks by integrating transcription factor-DNA interactions with expression data: an entropy-based approach. Computational Biology 19, 30–41 (2012)
9. Kleinberg, S., Mishra, B.: The temporal logic of causal structures. In: Proc. of the 25th Conf. on Uncertainty in Artificial Intelligence (UAI), pp. 303–312 (2009)
10. Rougny, A., Froidevaux, C., Yamamoto, Y., Inoue, K.: Translating the SBGN-AF language into logics to analyze signalling networks. In: Proc. of Int. Worksh. on LNMR, CORR, vol. 975, pp. 53–64 (2013)

11. Rougny, A., Froidevaux, C., Yamamoto, Y., Inoue, K.: Analyzing SBGN-AF networks using normal logic programs. In: Inoue, K., Farinas, L. (eds.) Logical Modeling of Biological Systems, IStE-Ltd. (to appear, 2014)
12. Le Novére, N., et al.: The systems biology graphical notation. Nature Biotechnology 27, 735–741 (2009)
13. Lewis, D.: Causation as influence. Philosophy 97, 182–197 (2000)
14. Mi, H., Schreiber, F., Le Novére, N., Moodie, S., Sorokin, A.: Systems biology graphical notation: activity flow language level 1. Nature Proceedings, 713 (2009)
15. Pearl, J.: Causality: Models, Reasoning, and Inference. Cambridge University Press (2000)
16. Muggleton, S.H.: Inverse entailment and Progol. New Generation Computing 13, 245–286 (1995)
17. Nabeshima, H., Iwanuma, K., Inoue, K., Ray, O.: SOLAR: An automated deduction system for consequence finding. AI Communications 23, 183–203 (2010)
18. Nakajima, N., Tamura, T., Yamanishi, Y., Horimoto, K., Akutsu, T.: Network completion using dynamic programming and least-squares fitting. The Scientific World Journal (2012), doi:10.1100/2012/957620
19. Nienhuys-Cheng, S.-H., de Wolf, R.: Foundations of Inductive Logic Programming. LNCS, vol. 1228. Springer, Heidelberg (1997)
20. Smet, R.D., Marchal, K.: Advantages and limitations of current network inference methods. Nature Reviews Microbiology 8, 717–729 (2010)
21. Tamaddoni-Nezhad, A., Chaleil, R., Muggleton, S.: Application of abductive ILP to learning metabolic network inhibition from temporal data. Machine Learning 65, 209–230 (2006)
22. Westergaard, S.L., Oliveira, A.P., Bro, C., Olsson, L., Nielsen, J.: A systems biology approach to study glucose repression in the yeast *Saccharomyces cerevisiae*. Biotechnology and Bioengineering 96, 134–145 (2007)
23. Whelan, K., Ray, O., King, R.D.: Representation, simulation, and hypothesis generation in graph and logical models of biological networks. In: Castrillo, J.I., Oliver, S.G. (eds.) Yeast Systems Biology, ch. 26, pp. 465–482 (2011)
24. Yamamoto, Y., Inoue, K., Doncescu, A.: Integrating abduction and induction in biological inference using CF-induction. In: Lodhi, H., Muggleton, S. (eds.) Elements of Computational Systems Biology, ch. 9, pp. 213–234 (2009)

Author Index